Model-Based Engineering with AADL

The SEI Series in Software Engineering

Software Engineering Institute | **Carnegie Mellon**

CERT Resilience Management Model

CMMI for Development

TSP

Documenting Software Architectures

Reflections on Management

↑Addison-Wesley

Visit **informit.com/sei** for a complete list of available products.

The **SEI Series in Software Engineering** represents is a collaborative undertaking of the Carnegie Mellon Software Engineering Institute (SEI) and Addison-Wesley to develop and publish books on software engineering and related topics. The common goal of the SEI and Addison-Wesley is to provide the most current information on these topics in a form that is easily usable by practitioners and students.

Books in the series describe frameworks, tools, methods, and technologies designed to help organizations, teams, and individuals improve their technical or management capabilities. Some books describe processes and practices for developing higher-quality software, acquiring programs for complex systems, or delivering services more effectively. Other books focus on software and system architecture and product-line development. Still others, from the SEI's CERT Program, describe technologies and practices needed to manage software and network security risk. These and all books in the series address critical problems in software engineering for which practical solutions are available.

Model-Based Engineering with AADL

An Introduction to the SAE Architecture Analysis & Design Language

Peter H. Feiler

David P. Gluch

⫩ Addison-Wesley

Upper Saddle River, NJ • Boston • Indianapolis • San Francisco
New York • Toronto • Montreal • London • Munich • Paris • Madrid
Capetown • Sydney • Tokyo • Singapore • Mexico City

Software Engineering Institute | **Carnegie Mellon**

The SEI Series in Software Engineering

The authors and publisher have taken care in the preparation of this book, but make no expressed or implied warranty of any kind and assume no responsibility for errors or omissions. No liability is assumed for incidental or consequential damages in connection with or arising out of the use of the information or programs contained herein.

The publisher offers excellent discounts on this book when ordered in quantity for bulk purchases or special sales, which may include electronic versions and/or custom covers and content particular to your business, training goals, marketing focus, and branding interests. For more information, please contact:

U.S. Corporate and Government Sales
(800) 382-3419
corpsales@pearsontechgroup.com

For sales outside the United States, please contact:

International Sales
international@pearsoned.com

Visit us on the Web: informit.com/aw

Cataloging-in-Publication data is on file with the Library of Congress.

ISBN-13: 978-0-13-420889-3
ISBN-10:   0-13-420889-7

This product is printed digitally on demand. This book is the paperback version of an original hardcover book.
First printing, September 2012

Contents

Preface

In this book, we introduce readers to the concepts, structure, and use of the SAE Architecture Analysis & Design Language (AADL) and demonstrate how AADL is an effective tool for Model-Based Engineering (MBE) of software system architectures. If you are just learning about AADL, we provide sufficient detail to enable you to develop and analyze basic system models. The core skills acquired by mastering the material in this book will provide you a foundation upon which to build your AADL and MBE expertise. Even when you are an accomplished AADL user, we anticipate you will find this book to be a valuable reference.

What and Why: MBE and AADL

Model-based engineering is the creation and analysis of models of your system such that you can predict and understand its capabilities and operational quality attributes (e.g., its performance, reliability, or security). By doing so throughout the lifecycle, you can discover system-level problems—those usually not found until system integration and acceptance testing—and avoid costly rework late in development and maintenance. In the past, separate models have been created for various system components and for each of the different analyses. A systematic and less fragmented approach is an architecture-centric one. Architecture-centric approaches address system-level issues and maintain a self-consistent set of analytical views of a system such that individual analyses retain their validity amidst architectural changes within the set.

The Architecture Analysis & Design Language (AADL) is an SAE International (formerly known as the Society of Automotive Engineers) standard [AS5506A[1]]. The AADL is a unifying framework for model-based software systems engineering that you use to capture the static modular software architecture, the runtime architecture in terms of communicating tasks, the computer platform architecture on which the software is deployed, and any physical system or environment with which the system interacts. You capture both the static structure and the dynamics in a single architecture model and annotate it with information that is relevant to the analysis of various operational characteristics. The concepts provided by AADL, such as threads, processes, or devices, have well-defined execution semantics that allow you to conduct both lightweight and formal analyses of systems. In addition, using its extensibility constructs, you as well as tool developers can blend custom analysis and specification techniques with core AADL capabilities to create a complete engineering environment for architectural modeling and analysis.

In developing an AADL model, you represent the architecture of your system as a hierarchy of interacting components. You organize interface specifications and implementation blueprints of software, hardware, and physical components into packages to support large-scale and team-based development.

As a standard, AADL provides you with the stability often not found in propriety technologies and allows you to participate in defining enhancements to the language. Additional elements of the standard suite that extend the AADL framework are found in the *SAE Architecture Analysis and Design Language (AADL) Annex Volume 1* [AS5506/1] and *SAE Architecture Analysis and Design Language (AADL) Annex Volume 2* [AS5506/2]. Released as a standard in June 2006, SAE AS-5506/1 defines annexes for the AADL graphical Notation, AADL Meta-Model and Interchange Formats, Language Compliance and Application Program Interface, and Error Model Language. Released as a standard in January 2011, SAE AS-5506/2 defines annexes for Behavior Modeling, for guidance on incorporating Data Modeling with AADL, and for ARINC653 Partitioned Architecture modeling.[2]

---

1. The standard AS5506A was originally published in November 2004. The book covers its revision published in January 2009, as well as errata corrections approved in 2012. For more information on the AADL, go to the Web site www.aadl.info. To purchase a copy of the standard, go to the Web site www.sae.org/technical/standards/AS5506A.

2. Additional annexes are in development for ballot in late 2012: a revision of the Error Model Annex standard, a Requirements Definition and Analysis Annex, and a Code Generation Annex.

Who Will Benefit from Reading This Book

You benefit from this book if you are a developer of software-reliant systems, whether a system or software architect, a system engineer, or an embedded software system developer. This book provides a foundation to enable you to apply the AADL and model-based engineering directly in your work. If you are a technical leader or project manager, the core principles and examples discussed in this book provide you with the knowledge required to guide technical personnel in the application of the AADL.

For graduate and advanced undergraduate software engineering students, this book offers a basis to understand and apply the AADL and MBE in your learning experiences. This book can be used as part of the material for a course on software architecture or software systems engineering of embedded real-time applications.

What You Need to Know to Get the Most Value from This Book

A basic knowledge of core software engineering practices (e.g., software architecture, software design), real-time systems (e.g., concurrency, scheduling, communications), and knowledge of computer runtime concepts (e.g., threads, execution semantics) will help you benefit most from this book. As a minimum, the level of expertise you should have in these areas is that commensurate with an advanced undergraduate student in computer science or software engineering. If you are a software developer with a degree in a technical discipline with two to three years' experience in developing embedded real-time software systems, you will find this book especially valuable in modeling software system architectures.

Structure of the Book

We have organized the material in the book into two parts, plus three appendixes. Part I is an overview of both the AADL language and MBE practices. It presents basic software systems modeling and analysis using the AADL in the context of an example system, including guidelines for effectively applying the AADL. Part II describes the characteristics of the elements of the AADL including representations, applicability, and constraints on their use. The appendixes include comprehensive listings of AADL language elements, properties that are defined as part of the AADL standard, a description of the example system used in the book, a list of references, and an index.

Terminology

AADL is a component-based modeling language that distinguishes between component interface specifications (component type declarations), component implementation blueprints (component implementation declarations), and component instances (subcomponent declarations). Component types and implementations are referred to as component classifiers. AADL also distinguishes between component categories with specific semantics to model the application software (e.g., thread, process, data), the execution platform (e.g., processor, bus, device), and composite components (system). The AADL standard document uses terms such as *system type declaration* or *system implementation declaration*. In this book, we use abbreviated terms such as *system type* or *system* where the context makes the meaning clear.

Example Application System

We use a powerboat autopilot (PBA) control system as the basis for most of the examples throughout this book. The PBA is an embedded real-time system for the speed, navigational, and guidance control of a maritime vessel. However, the PBA is an invention created to provide a backdrop for demonstrating the AADL and does not represent any specific commercial, military, or research system. While the PBA is a maritime application, it represents key elements of vehicle control for a wide range of applications including aircraft, spacecraft, and automotive and other land vehicles.[3]

---

3. Details of the PBA system are provided in Appendix A.

About the Authors

Dr. Peter Feiler is a Senior Member of Technical Staff in the Research Technology and Systems Solutions (RTSS) program at the Software Engineering Institute (SEI). He is a 27-year veteran of the SEI. His interests include architecture-centric engineering of safety-critical embedded real-time systems. He is collaborating with researchers at Carnegie Mellon University and other research institutions to develop model-based architecture technology and is investigating its practicality with commercial industry. He is the author and editor of the SAE International (formerly known as Society of Automotive Engineers) Architecture Analysis & Design Language (AADL) standard. Peter has a Ph.D. in computer science from Carnegie Mellon University and is a senior member and member of ACM, IEEE, and SAE International. He recently received the Carnegie Science Award for Information Technology.

   Dr. David P. Gluch is a professor in the Department of Electrical, Computer, Software, and Systems Engineering at Embry-Riddle Aeronautical University and a visiting scientist at the Software Engineering Institute (SEI). His research interests are technologies and practices for model-based software engineering of complex systems, with a focus on software verification. Prior to joining the faculty at Embry-Riddle, he was a senior member of the technical staff at the SEI where he participated in the development and transition of innovative software engineering practices and technologies. His industrial research and development experience has included fault-tolerant computer, fly-by-wire aircraft control, Space Shuttle software modeling, and automated process control systems. He has co-authored a book on real-time UNIX systems and authored numerous technical reports and professional articles. Dave has a Ph.D. in physics from Florida State University and is a senior member of IEEE.

Acknowledgments

We would like to thank a number of people for helping make this book a reality.

   We would like to thank Bruce Lewis as the chair of the SAE AADL committee in making AADL a reality. The quarterly standards meetings provided a forum for user feedback on the use of AADL. We would also like to thank the members of the committee, especially those from industry, in helping to shape AADL into a language that meets a practical need. It was in this setting that the idea for a book on the use of AADL in model-based engineering came about.

We also appreciate the efforts of the research and advanced technology community from various universities and industry in using AADL as a platform for a wide range of formal software-reliant system analysis and demonstrating the feasibility of model-based engineering with AADL. Their tools and technology to drive the analysis of architectures allows AADL to show off its strength.

At the Software Engineering Institute (SEI), we are thankful to Tricia Oberndorf and Linda Northrop, our program managers for allowing us to invest time and effort into the endeavor of writing this book and encouraging us to bring it to completion. The other SEI AADL team members, Lutz Wrage, Aaron Greenhouse, John Hudak, Joseph Seibel, Dio DeNiz, and Craig Meyers, contributed in various form to the body of knowledge on the use of AADL, a small portion of which is reflected in this book. They led and contributed to the creation and use of the OSATE tool set, the development and presentation of tutorials and two courses on AADL, and the use of AADL on customer projects. We also received feedback on various drafts of the book.

We appreciate the feedback from external reviewers of book drafts, in particular Bruce Lewis, Jérôme Hugues, and Oleg Sokolsky. Finally, we want to thank Peter Gordon and Kim Boedigheimer from Addison-Wesley for the production of the book.

Introduction

As embedded systems have become increasingly software intensive and dependent upon commercial computing hardware, system integration has emerged as a major engineering challenge in the development of modern application systems. Model-based engineering practices that include systematic analyses of system architecture models early and throughout the development lifecycle can provide substantially greater confidence that the integrated system will meet design goals.

Early in the lifecycle, model-based analysis takes the form of predictive analysis, for example addressing uncertainty through sensitivity analysis. As a system architecture is refined, the model includes finer granularity and analysis of that model provides more detailed insights into the qualities and capabilities of the design. As system components are developed and measurements are taken of the implementation, the process increasingly takes on the character of validation by comparing and replacing predictions with actual measurements. A key to these practices is the analysis of a system's operational characteristics using models of its software runtime architecture, its compute platform, its interface to the physical environment, and the deployment of software onto hardware. Cumulatively, the insights from these analyses help ensure a more efficient, less error prone system development and system integration process and provide greater assurance that the delivered system will meet customer expectations.

The Architecture Analysis & Design Language (AADL) provides a rigorous and extensible foundation for model-based engineering analysis practices that extend throughout system design, integration, and assurance. With the AADL, you represent systems as distinct software and hardware components and their interactions. By relying upon its formal foundations, you conduct both lightweight and rigorous analyses of critical real-time computational factors such as performance, dependability, security, and data integrity. In addition, with the AADL extensibility constructs, you can integrate additional established and custom analysis and specification techniques into an engineering environment that allows you to develop and analyze a single, unified system architecture model.

Part I

Model-Based Engineering and the AADL

Engineers create models that range from cryptic sketches on the back of a dinner napkin to formal mathematical representations, often augmented by full-scale physical simulations. Collectively, these representations not only capture and provide insight into a design but formal models enable the analysis and prediction of a product's critical aspects before it is operational. As technology has advanced, the sophistication of models and their role has expanded until models are indispensable in the engineering of modern application systems. For example, the 777 "is the first Boeing airliner 100% designed using 3-D solid modeling technology" and "… was the first 100% digitally designed and pre-assembled airplane made by Boeing." [BOE]

   AADL was designed to provide such a modeling and analysis capability for engineering software systems. These models are architecture models of a system that consist of software, hardware, and physical system components, their interactions, and the properties of these elements. The language concepts for modeling system architectures have well-defined semantics that allow you to drive analyses of different critical system factors from the same architecture model source. In addition, through AADL extensibility constructs, custom analysis and specification techniques can be combined with core AADL capabilities to create a complete engineering environment for architectural modeling and analysis.

While our emphasis is on architectural modeling and analysis, the AADL supports the generation of application specific runtime systems and the auto-build of complete systems from validated AADL models. These capabilities can be included in an architecture-centric development environment that spans requirement analysis, architectural and detailed design, as well as validation and code generation.

In Part I, we provide guidance and representative examples of applying the AADL in system architectural design and analysis within a model-based software systems engineering environment. Initially, we present an overview of model-based engineering and the AADL. In Chapters 2 and 3, we discuss the basics of using the AADL and provide an example of its use in modeling and analyzing a simple control application. In Chapter 4, we provide additional details on the use of many of the AADL capabilities for architectural modeling and analysis.

Chapter 1

Model-Based Software Systems Engineering

The phrase *model-based engineering (MBE)* is used to designate engineering practices in which models are the central and indispensable artifacts throughout a product's lifecycle encompassing concept, development, deployment, operation, and maintenance. Increasingly, this phrase is being applied to software engineering to emphasize its maturation as an engineering discipline, one where modeling technologies are becoming indispensable in developing advanced computing systems. MBE has emerged within software engineering in a variety of guises including model-driven architecture (MDA), model-driven development (MDD), model-based development (MBD), and model-centered development (MCD). Architecture descriptions are a key element of the IEEE standard *Systems and Software Engineering—Architecture Descriptions* [IEEE 42010]. This section introduces the AADL as a technical foundation for architecture-centric model-based software systems engineering and discusses its relationship to other software engineering practices and techniques.

1.1 MBE and Software System Engineering

Abstraction and encapsulation are key principles for managing the complexity of software. You can find these in the class/method concept of object-oriented design, and the module/package concept in both programming and modeling languages. The focus of the package concept is on structuring software modeling, design, and program artifacts by limiting their visibility and use in order to improve modifiability of the software.

Architecture languages have introduced the concept of component to structure the architecture of a system. Simulink blocks do this for control systems, SysML blocks for physical systems, UML for conceptual architecture, and AADL for the architecture of embedded systems.

Because software requires hardware for its execution, effective software MBE practices for computing systems must not only consider the static architecture of the application software, but also address the operational dynamics of the runtime architecture and its deployment on the computer platform (e.g., how long the software takes to execute on a specific processor). This is especially the case for embedded application systems. In these systems, software is an integral part of an application platform (e.g., aircraft, automobiles, spacecraft) and its interfaces to the operating environment. Often these software-dependent systems control the behavior of other systems in that environment. For instance, software controls the inflation of automobile airbags during a crash.

1.1.1 MBE for Embedded Real-Time Systems

In this book, we use MBE to designate model-based engineering of real-time embedded software-intensive application systems. These are computational systems within an operational environment that provide monitoring, control, or other critical functions. In these application systems, performance, safety, and dependability are vitally important factors (i.e., vitally important operational quality attributes). Systems in the automotive, avionics, aerospace, medical, robotics, and process control industries among others fall into this category. Examples of these attributes include the reaction time of the guidance software controlling a missile or the probability of failure of a multiply redundant flight control system in an aircraft.

Figure 1-1 illustrates the three key elements of an embedded software system architecture:

- A representation of the application runtime architecture in terms of communicating application tasks
- A representation of the computer platform in terms of its processors and memory interconnected by networks
- A representation of the physical system and/or environment with which the embedded application interacts

Not only are these three elements essential but also their interactions in the form of

- Logical connectivity between the embedded application software and the physical system components in the form of observations, measurements, control, and commands
- Physical connectivity between the computer system and the physical system through its sensors and actuators
- Deployment of the application software tasks onto the computer platform to utilize its computational resources for its execution

The logical connection between the embedded application and the physical system allows us to ensure that the embedded application

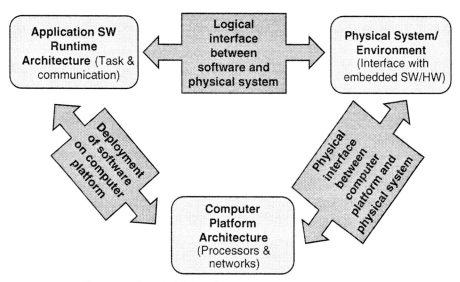

Figure 1-1: *Elements of Embedded Software System Architectures*

makes the right assumptions about the physical system and its behavior including faults and exceptional behavior. The physical connection between the computer system and the physical system allows us to ensure that they can physically interact and system engineering requirements such as electrical power and mass are satisfied. The deployment of the software on the computer hardware allows us to ensure that these resources are shared properly and the performance and safety criteria assumed by the system engineer and the control engineer are met.

In this computer systems context, MBE models represent software and computational hardware resources, and the interfaces to the physical system under control. Models of the physical characteristics of the system under control may not be included explicitly. For example, in an avionics application the aerodynamics of aircraft wings may not be explicitly represented in an AADL model of the flight control system. However, since the aircraft is being controlled by this system, interfaces to aircraft components (e.g., to an aircraft's aileron position sensors) and explicit models and properties of the hardware needed for physical sensing and control (e.g., an aircraft's spoiler actuator, properties of the aircraft's rudder) appear in a comprehensive AADL model of an aircraft's control system.

1.1.2 Analyzable Models and MBE

Analyzable models provide a foundation for prediction and design assurance by enabling meaningful assessment and reasoning about their properties and the properties of the systems they represent. Model properties are addressed in the development process and focus on quality attributes such as modifiability and maintainability, while system properties address operational quality attributes such as performance, safety, and reliability. These models range from notations with minimal syntactic and semantic rules that can be scrutinized for compliance and correctness to quantitatively based and rigorously detailed representations for which precise simulations and extensive automated analyses can be readily accomplished.

For example, consider a simple block diagram that includes a legend defining the notation used in the diagram. If the legend simply identifies rectangles as objects and interconnections as data flows but does not describe the computational significance, execution timing, or runtime characteristics of those elements, the representation is limited principally to manual inspection. It is possible to check that all rectangles are connected by data flows and the directions of the flows are consistent. However, assessing the data consistency within flows, system performance, and transport times for data through the system is not readily achievable.

In contrast, a model with precise descriptions of the runtime nature and timing semantics of its elements (e.g., paths of concurrent execution, execution frequencies, and worst-case execution times) enables rigorous quantitative analyses. Analyses that can be readily performed include automated data consistency checks, rate-monotonic analysis, performance assessments, and detailed system simulation. Other formal representations such as state machine and temporal logic models enable theorem proving and symbolic model checking of the system's behavior.

Despite the benefit of analytical models, in particular models that have a strong formal foundation, there are pitfalls in their use. Companies have experienced system failures despite the fact that their analytical models told them that everything was fine. These analytical models are often developed by separate teams (e.g., a timing model by the performance team, a security model by the security team, and a fault model by the reliability team). They may develop their models at different points in the development life cycle based on a written document while the system architecture evolves. As a result the models are not consistent with each other and do not reflect the architecture of the system that is to be deployed, as illustrated in Figure 1-2.

You can address this problem with an architecture-centric approach to MBE. You use an architecture model as the single source to drive different dimensions of analysis. You do so by annotating the architecture model with information that is relevant to the automatic generation of analytical models for various quality attributes that are critical to your system, as illustrated in Figure 1-3.

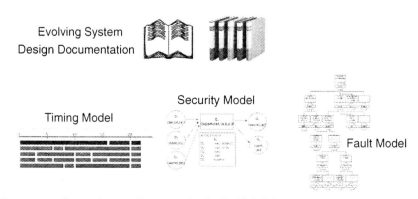

Figure 1-2: *Inconsistency Between Analytical Models*

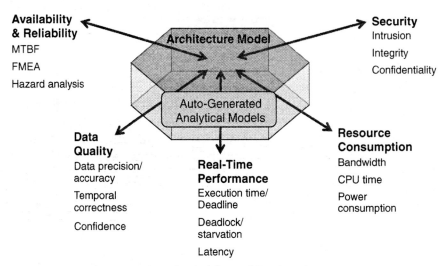

Figure 1-3: *Architecture-Centric Model-Based Engineering*

This architecture-centric approach has two major benefits over traditional MBE approaches. The first benefit is that any changes to the architecture itself, such as addition or replacement of a system component or an interaction between components, is automatically reflected in the analytical models when they are regenerated to revalidate the system after an architecture change. The second benefit is that the impact of a change to the system to address one quality attribute (e.g., a larger bit encryption algorithm to improve security) on other quality attributes (e.g., potential problems meeting timing requirements due to increased computation) can be easily determined by re-evaluating the analytical models of different quality dimensions.

1.1.3 MBE and the AADL

The need for analyzable architecture models was recognized in the 1990s by the aerospace industry due to the increasing role of embedded software in system integration as it was embracing the concept of integrated modular avionics (IMA). This led to the development of AADL starting in 1999. The language is patterned after MetaH, a research prototype of a language and tool suite for analysis and generation of embedded real-time systems, which was developed by Honeywell Technology Center and successfully deployed in various pilot projects throughout the 1990s. AADL was originally published in November

2004 and has been used in a number of industrial and research initiatives. Based on feedback from that experience AADL has been improved and the revised standard became available in January 2009.

AADL is both a textual and graphical language with component-based modeling concepts specifically designed to represent embedded software systems. AADL provides data and subprogram components organized into packages to abstractly represent application source code that is implemented in any programming language (such as Java, C, or Ada) or in an application design language (such as Simulink for control system components). AADL provides threads, thread groups, and processes to represent concurrent tasks executing in protected address spaces (time and space partitioning) and interacting through ports, shared data components, and service calls to represent the software runtime architecture. The dynamics of the runtime architecture are captured through mode state machines at different levels of the component hierarchy to represent operational modes, dynamic changes to fault-tolerant configurations, and component behavior. The AADL standard provides well-defined execution semantics for task execution, communication timing, and mode changes using hybrid automata specifications to address predictable response times.

AADL provides processor, memory, and bus components to represent computer platform architectures in terms of hardware, and the virtual processor and virtual bus components to represent virtual machines, partitions, virtual channels, and protocols as runtime system abstractions. Modeling of the physical system is supported through the device and bus concepts. The system component is used to represent composite components of the application architecture, the computer platform, and the physical system. The abstract component and parameterization of component specifications supports component templates and architecture patterns.

AADL is extensible to support annotation of models with user-defined and analysis-specific properties and with specialized annex sublanguages. The Behavior Annex standard for AADL [BAnnex] extends the core AADL to specify the component interaction behavior with further precision to address safety-criticality aspects of the system. The Error Model Annex standard for AADL [EAnnex] extends the core AADL to specify fault behavior and error propagation to address reliability aspects of the system architecture. Other annexes are in development.

In addition, the concepts of AADL have been reflected in a UML profile for embedded systems. This profile is part of the OMG effort to provide a UML profile for Modeling and Analysis of Real-time and Embedded systems (MARTE) [MARTE].

The AADL standard includes an XML interchange format defined in terms of a Meta model for AADL (XMI) to facilitate interchange of models for virtual system integration, to automatically transform annotated AADL models into analytical models, and the ability to interface commercial, in-house, and research tools into an embedded systems engineering environment. This allows companies to adopt AADL into their existing environment and integrate with their established tool base to migrate to a scalable and extensible architecture-centric MBE solution that addresses their modeling, analysis, and auto-generation needs.

1.2 AADL and Other Modeling Languages

Model-based techniques are ubiquitous in the software and systems engineering community. Mathworks Simulink is a model-based design and simulation environment for dynamic control and embedded systems. Its block libraries support modeling of continuous time and discrete time designs into a component hierarchy. Simulink models represent control functionality of application components that are associated with AADL threads and processes in the same way as source text written in programming languages. Computer hardware engineers use the IEEE standard VHDL (Very High Speed Integrated Circuit (VHSIC) Hardware Description Language) for detailed design and architecture block notations with roots in the Processor Memory Switch (PMS) used the 1971 Bell and Newell classic book *Computer Architectures: Readings and Examples*. AADL supports the PMS paradigm through processor, memory, and bus components with VHDL models associated as source text. Modelica is an object-oriented equation-based language to model physical systems that contain mechanical, electrical, hydraulic, thermal, and process-oriented components.

The Object Management Group (OMG) has embraced the model-based engineering paradigm through Model-Driven Architecture (MDA), the Unified Modeling Language (UML) [UML], and UML profiles such as System Modeling Language (SysML) [SysML], and Modeling and Analysis of Real-Time Embedded systems (MARTE). UML started out as a detailed design notation for object-based software and expanded with UML2 to support component-based modeling.

SysML was introduced as a layer of modeling concepts on top of UML for the specification of complex systems by system engineers. Such extensions are known as UML profiles. Recently, the MARTE profile has been defined for embedded software systems supporting the creation of AADL models.

This leads us to an MBE practice that spans system engineering, embedded software system engineering, hardware engineering, control engineering, application software development, and other engineering disciplines such as mechanical engineering. Such an MBE practice requires architecture-centric co-engineering of systems through a combination of modeling capabilities to predictably meet safety criticality, performance, reliability, security, and other operational quality attributes.

Figure 1-4 shows the three elements of an embedded system architecture, namely the embedded application software and its runtime system, the computer platform, and the physical system they interface with. It also shows the detailed design of components in each, as well as the operational environment. The focus of system engineering is on the system as a whole in its operational environment, defining the requirements, structure, behavior, and parametrics of the system and its components in an architecture notation such as SysML, with the

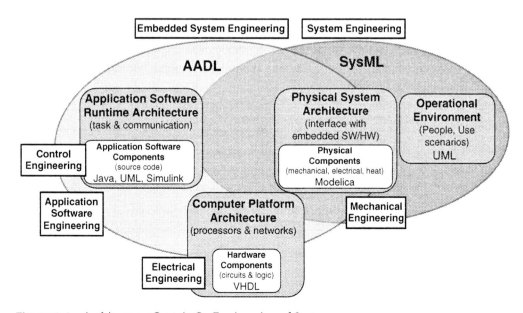

Figure 1-4: *Architecture-Centric Co-Engineering of Systems*

computer platform as one of the components and the system function-ality as a specification of the application software. Embedded software system engineering focuses on the embedded software runtime archi-tecture, and its interaction with the computer platform, and the physi-cal system. The figure shows the refinement of component designs in each of these architecture elements in specialized notations. Physical component models are expressed in notations like Modelica or Simulink to capture mechanical, electrical, fluid dynamic, and other physical aspects. Computer hardware models are expressed in notations like VHDL to capture the electronic circuit logic design of hardware. Application software components are designed and implemented using programming languages or detailed design notations such as Simulink for control system design or UML for general software design.

The remainder of this section elaborates on the interaction between AADL and MDA/UML as well as the relationship between AADL and SysML.

1.2.1 AADL, MDA, and UML

The Model Driven Architecture (MDA) is a comprehensive approach to software system development that supports the entire product lifecy-cle, addressing management, business, and technical aspects. It relies on UML and related technologies and practices. One of the central con-cepts in the MDA is the partitioning of models into platform indepen-dent models (PIM) and platform specific models (PSM). This separation of concerns pervades the approach and allows established domain spe-cific models to be used as the basis for specialized implementations employing new technologies [MDA].

The AADL is effective for the development of architecture models with platform independent elements and platform elements, allowing extensive quantitative analysis early and throughout the life cycle that capitalizes on the well-defined semantics of the language. Without platform specific information, analysis of operational quality attributes would not be possible. These capabilities support technical risk mitiga-tion through early discovery of system-level problems.

UML is a "graphical language for visualizing, specifying, construct-ing, and documenting the artifacts of distributed object systems" [UML]. UML's genesis was as a modeling language for detailed design of software, emphasizing an object-oriented approach. However, its applicability extends beyond software systems. The UML has evolved to include specialized constructs to facilitate software architectural and component-based modeling. In version UML 2.0, constructs are included to enable component-based design, generalized interfaces, and hierarchical representations.

UML supports the software development process through modeling of the context and intended use of an application captured in use case scenarios, data modeling through class diagrams, representation of a conceptual architecture through a combination of classes and components, behavior modeling through state charts, and modeling of interactions between multiple active components with activity and sequence diagrams. As such, UML is able to address the platform independent modeling (PIM) aspect of MDA, but industrial users have recognized its shortcomings in addressing the non-functional characteristics of performance-critical (embedded) systems in platform specific models (PSM). AADL confronts this issue by directly addressing the performance-critical aspects of software system architecture.

1.2.2 AADL and SysML

SysML is "a general-purpose (graphical) modeling language for the specification, analysis, design, verification and validation of complex systems including hardware, software, information, processes, personnel, and facilities" [SysML]. It includes considerations of application domain as well as computational resources to support application functionality. SysML is a general-purpose graphical modeling language developed for system engineering that is a subset of UML 2.0. A SysML model is a collection of diagrams that describe the system's components and their interactions to express the requirements, the structure, the behavior, and the equation-based parametrics of systems.

There are notable differences in the foundations for the AADL and SysML. The AADL is a rigorous textual modeling language with graphical portrayals of modeling elements [AS5506A]. SysML is a graphical language for effective visual presentation that is based upon UML. It incorporates two new diagrams applicable to systems engineering that are not found in UML: the requirements and the parametric diagram. The requirements diagram supports more structured requirements documentation and facilitates traceability. The parametric diagram captures the dynamics of physical systems in mathematically precise detail that enables simulation of the dynamics. However, these diagrams are less effective in capturing and representing execution of a computer's runtime environment that includes threads, processes, and their allocation to different processors. The rest of SysML lacks a defined foundation for rigorous formal analyses, as noted in [SysML]: "SysML is specified using a combination of UML modeling techniques and precise natural language to balance rigor and understandability. Use of more formal constraints and semantics may be applied in future versions to further increase the precision of the language."

The relationship of SysML and the AADL in capturing the architecture of an embedded system is shown in Figure 1-4. SysML focuses on the system as a whole in the context of its operational environment with the computer platform as one component and the software implementing system functionality. The AADL focuses on the interaction between the physical system architecture, the runtime architecture of the embedded application software, and the computer platform. It provides insights into the impact of implementing system functionality in software and decisions made in the application runtime architecture, runtime infrastructure such as operating system and communication protocols, and the deployment on a distributed computer platform. Such a focus is critical to validating assumptions made by system engineers and control engineers about the management and control of a system in terms of timing, reliability, and fail-safe behavior of the embedded software system. Catastrophic failures such as the self-destruction of Ariane 5 [Ariane 5] have shown the importance of such an engineering focus to complement system engineering.

Chapter 2

Working with the SAE AADL

With the AADL, you capture the architecture of embedded systems as architectural models that provide well-defined analyzable semantics of its runtime architecture. The creation of these descriptions encompasses identifying and detailing software and hardware components and their interactions in the form of interface specifications (component types) and blueprints of their implementation (component implementations) and organizing them into packages. These packages effectively represent libraries of component specifications that can be used in multiple architecture models, can be authored and maintained by different modelers, and may be version controlled. Packages have public and private sections in support of information hiding. Public package sections contain specifications available to other packages. You use the private package section to protect details of component implementations you do not want to expose to others. AADL models may also include property sets through which you introduce new user-defined properties to the AADL model.

For software components such as threads, processes, and data, you define their runtime characteristics including execution times, scheduling, and deadlines. In addition, you declare other code related information such as execution code size, source code language, source code file names, and source code size.

> Threads (execution paths), data, and processes (representations of compiled code and data) are the principal runtime software abstractions within the AADL.

In describing hardware, you define execution platform components such as processors, memory, and buses, and characterize them with relevant execution and configuration properties. For example, in defining the execution hardware for a system, you declare the processors that execute threads, memory that stores code and data, and buses that provide physical connections between hardware components. The values of properties associated with these components are used to define characteristics such as processor speed, memory word size, etc. In addition, you can define devices to represent interfaces to a system's operational environment such as sensors and actuators.

> Processors, memory, buses, and devices are the hardware abstractions within the AADL. They provide the execution environments, physical communication paths, and external interfaces for a computing system.

To complete the architectural description, you define a fully specified system implementation by integrating all of a system's composite elements into a single component hierarchy. This hierarchy establishes all of the interactions among the components and the architectural structure required to define an executable system. These encompass data and event exchanges and the physical connections among components as well as the assignment of software to hardware (e.g., where data and execution instances are stored, where software executes, and the hardware supports data and event transfer). When you instantiate this top-level system implementation you get an AADL instance model that is a mirror image of the system you are modeling. This instance model is the basis for analyzing operational properties of your system. Throughout the development of an architectural model, you can conduct extensive analyses of consistency and compliance with requirements. These analyses range from syntactic compliance and basic interface data consistency checking to rigorous assessments of quality attributes and behavior. For example, using AADL formal representations and specialized extensions, you can investigate properties such as data latency, bandwidth, scheduling, and dependability.

2.1 AADL Models

Central to an AADL model are component type and implementation declarations, organized into packages (see Section 4.4.1). In a component type declaration, you define the category and interfaces (features) of the component. This corresponds to a specification sheet for a component. In a component implementation declaration of that type, you define the internal structure of the component (i.e., its constituents and their interactions). This corresponds to a blueprint for building a component from its parts. In these declarations, you specify the properties required to define fully a component's runtime characteristics. These properties may be one of the predeclared properties in the AADL standard, properties introduced through standard annexes for specific analyses, or user-defined properties that you have introduced through additional property sets. Collectively, these declarations (a type plus an implementation) define a pattern for a component. Component types and implementations are referred to as component classifiers.

2.1.1 Component Categories

Components are grouped into application software, execution platform, composite, or generic (non-runtime specific) components. These are summarized in Table 2-1.

Table 2-1: *Component Categories*

Application Software	`data`	Data in source code and application data types
	`thread`	A schedulable unit of concurrent execution
	`thread group`	An abstraction for logically organizing threads, thread groups, and data components within a process
	`process`	Protected address space enforced at runtime
	`subprogram`	Callable sequentially executable code that represents concepts such as call-return and calls-on methods
	`subprogram group`	An abstraction for organizing subprograms into libraries

continues

Table 2-1: *Component Categories (continued)*

	`processor`	Schedules and executes threads and virtual processors
	`virtual processor`	Logical resource that is capable of scheduling and executing threads that must be bound to or be a subcomponent of one or more physical processors
Execution Platform (Hardware)	`memory`	Stores code and data
	`bus`	Interconnects processors, memory, and devices
	`virtual bus`	Represents a communication abstraction such as a virtual channel or communication protocol
	`device`	Represents sensors, actuators, or other components that interface with the external environment
Composite	`system`	Integrates software, hardware, and other system components into a distinct unit within an architecture
Generic	`abstract`	Defines a runtime neutral (conceptual) component that can be refined into another component category

2.1.2 Language Syntax

The AADL is a formal declarative language described by a context-free syntax [AS5506A].[1] In the language, names and reserved words are case insensitive. In our examples, reserved words are shown in lower-case boldface type by convention. Listing 2-1 shows an example AADL declaration for a package *controllers* containing a process type declaration in its **public** section. There are two **features** defined for the **process** *speed_control*: an **in data port** and an **out data port**. These **features** define data input and data output interfaces for the process type. An **end** statement terminates the package and process declarations. Blank lines shown in the example are optional. They are included to improve readability.

---

1. Listing A-1 in Appendix A summarizes the AADL syntax and grammar rules.

Listing 2-1: *Example AADL Declarations*

```
package controllers
public

process control
features
  input_speed: in data port;
  output_cmd: out data port;
end control;

end controllers;
```

2.1.3 AADL Classifiers

Component types and implementation declarations are classifiers. You reference classifiers by their name to establish an occurrence (an instance) of the pattern that they describe. These references are made within individual textual statements contained within type and implementation declarations. For example, within a type declaration a single textual declaration is used to declare that a component has an in data port with the name *input_speed* and the type of the data that passes into the component through the port is identified by referencing a data type (e.g., *speed_data*). A sample component type declaration is shown in Listing 2-2, which includes an **in data port** declaration that references *speed_data*.

Listing 2-2: *Sample Component Type Declaration*

```
process control
features
  input_speed: in data port speed_data;
  toggle_mode: in event port;
  throttle_cmd: out data port throttle_data;
  error_set: feature group all_errors;
flows
  speed_signal_path: flow path input_speed -> throttle_cmd ;
properties
  Period => 20 ms;
end control;
```

As another example, consider the subcomponent declarations within the partial implementation declaration shown in Listing 2-3. Note that this is an implementation of the process type *control* as reflected in its name *control.speed*. In this example, the subcomponent *read_speed* is declared as an occurrence (instance) of the thread component implementation (pattern) *read_data.speed* and the subcomponent *control_laws* is declared as an instance of the thread component implementation *control.basic*.

Listing 2-3: *Sample Component Implementation Declaration*

```
process implementation control.speed
subcomponents
  read_speed: thread read_data.speed;
  control_laws: thread control.basic;
connections
  c1: port input_speed -> read_speed.in_data;
  c2: port read_speed.out_data -> control_laws.in_data;
  c3: port control_laws.cmd -> throttle_cmd;
flows
  speed_signal_path: flow path
    input_speed -> c1 -> read_speed -> c2 ->
    control_laws -> c3 -> throttle_cmd ;
end control.speed;
```

2.1.4 Summary of AADL Declarations

Table 2-2 summarizes the set of AADL declarations. The left column lists classifier (component type, component implementation, and feature group declarations), package, and property set declarations and a simplified example of each. The right column summarizes the structure and content of the declarations listed in the left column.

You can include up to seven sections within a component type declaration. These sections are shown in the order in which they are declared. One section is **features**, the section in which interfaces (e.g., ports) of a component are declared. The **flows** section allows you to talk about flow through a component without exposing the component implementation and attach properties such as latency or computational error to the flow. The **modes** section allows you to talk about operational modes of the component. The **prototypes** section allows you to parameterize the component type as a template and you use the **extends** construct to refine the component type template. You may also include a **properties** section within a type or implementation declaration. The properties section is where values are declared for properties associated with a component or feature group. You may also adorn many of the declarations listed in the column on the right (i.e., features, flows, modes) with property values. For example, when you declare a thread

subcomponent you can assign a specific value of the Period property
for that thread subcomponent. Annex library and annex subclause declarations enable you to extend the capabilities of the AADL language.

Table 2-2: *AADL Declarations*

Declaration	Content

Component Type
 	
`system control`
`<component type content>`
`end control;`
	• Extends
	• Prototypes
	• Features
	• Flows (specifications)
	• Modes
	• Properties
	• Annex subclauses

Component Implementation
 	
`system implementation control.flight`
`<component implementation content>`
`end control.flight;`
	• Extends
	• Prototypes
	• Subcomponents
	• Connections
	• Calls
	• Flows (implementations)
	• Modes
	• Properties
	• Annex subclauses

Feature Group Type
 	
`feature group sensor_array`
`<feature group content>`
`end sensor_array;`
	• Extends
	• Prototypes
	• Features
	• Properties

Package
 	
`package control_hardware`
`<package content>`
`end control_hardware;`
	• Public
	    – Visibility
	    – Classifier
	    – Annex library
	• Private
	    – Visibility
	    – Classifier
	    – Annex library

Property Set
 	
`property set fault_properties is`
`<property set content>`
`end fault_properties;`
	• Property type
	• Property name
	• Property constant

Component implementations consist of up to nine sections. The extends, prototypes, properties, and annex subclauses sections have the same role as for component types. You use the subcomponents section to specify the composition of the component from instances of subcomponents of various kinds by referring to their classifier. In the connection section, you indicate how subcomponents interact with each other and, through the features of the component type, with external components. The calls section outlines call sequences to subprograms from inside thread and subprogram implementations. The flows section details how a flow specification in the component type is realized as a flow through the subcomponents. The modes section may add mode transition information to modes specified in the component type or document component implementation specific modes.

In addition to component classifiers, there is a feature group type classifier. A feature group type defines a collection of component features. You indicate that a collection of connections may exist between two subsystems by grouping the features into a feature group for each and connecting them with a single connection. You can do so early in the modeling process as you define the subsystem architecture without detailing the individual features, and later provide the feature details by filling in the feature group type. You reference a feature group type when you declare a feature group within a component type declaration. This is shown in Listing 2-2, where the feature group *error_set* references the feature group type *all_errors*.

Packages allow you to organize component types, component implementations, and feature group types into a named group type. Packages represent separate namespaces for their content such that a component type is identified by its package name and its component type name (e.g., the component type *pitch_control* that is declared in a package *controllers* is referenced as *controllers::pitch_control*). One objective of grouping component specifications into a package is to allow different teams to develop a system, each team working on different parts independently. In this situation, a package may be used as a unit of version control.

A property set enables you to define new properties and property types in a separate namespace. These properties are identified by their property set name and property or property type name. This is especially valuable in customizing a description and in extending analyses (e.g., adding a specialized security level property to conduct a system-level security analysis). Property sets are not declared inside packages, but separately as part of an AADL model.

2.1.5 Structure of AADL Models

You distinguish between the structure of a system architecture and the structure of an AADL model. You use a hierarchy of interacting components to represent the system architecture. You use packages to organize component interface specifications (component types) and their blueprints (component implementations) into libraries such that a collaborating team can evolve the architecture model. You introduce new properties by grouping them into different property sets. In other words, an AADL model consists of packages and property sets. All AADL declarations must be contained within a package or property set declaration.

Package and property set declarations cannot be nested in any way (i.e., a package cannot be declared within another package or property set nor can a property set be declared within another property set or in a package). However, you can nest package names, similar to packages in the Java language. For example you can declare a package called *sei::boats::gps.*

Figure 2-1 presents the structure of AADL models. In the figure, rectangles represent declarations. The dashed arrow between a type and an implementation declaration indicates that the implementation belongs to the type pointed to by the arrow. The asterisk adorning the rectangles and the implementation links indicates zero or more occurrences. There may be multiple implementations associated with a single type and there may be types without implementations. In addition,

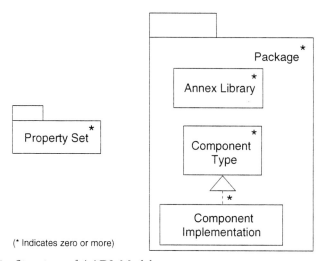

Figure 2-1: *Structure of AADL Models*

while there may be multiple annex library declarations within a single package, there must only be one annex library declaration for each annex.

You can have a complete AADL model contained within a single file or have packages and property sets grouped into multiple files, even to the extent that there is a separate file for each package and each property set. These files can be worked on separately by different team members and can be version controlled. The partitioning of a specification into files is tool specific. For example, while not required by the standard, the Open Source AADL Tool Environment (OSATE) [OSATE] assumes that each package and property set is in a separate file. The AADL standard allows the AADL models to be stored in a database instead of a file system. In that case, there is no correspondence of packages to files.

2.2 System Specification and System Instances

A collection of AADL packages and optional property sets represent a system specification in AADL, referred to as AADL declarative model. It represents an architectural blueprint that describes the structure, properties, and connectivity of an embedded system.

A system instance model (or simply an instance model) is an instantiation of a system model using a system component implementation with a declarative system model as its root. You can create instance models of the whole system or of a particular subsystem of interest by choosing the appropriate system implementation declaration as the starting point. The component implementations of the declarative model are interpreted as blueprints to create and interconnect all the component instances of the system.

You use this instance model to analyze your system architecture and to drive the generation of its implementation using analysis capabilities built into or as extension of toolsets like OSATE, or through separate analysis via a standardized interchange representation. For additional information on model creation and validation tools, please see Chapter 15.

2.2.1 Creating System Instance Models

A system instance model represents an operational physical system that is generated from a fully specified declarative model. In a fully specified declarative model, the runtime architecture is represented to the level of threads and subprogram calls with application software components modeled as subprograms and data component types. In addition, a fully specified system includes processors to execute application code, memory to store application code and data, devices that represent the physical environment of the embedded application, and buses that connect these components.

AADL allows you to model your system to the level of fidelity appropriate for the development phase and analyses of interest. In other words, you may have created a partially complete model of the system in terms of major subsystems, or in terms of processes with or without threads. You can create an instance of such a partial model and analyze it. AADL allows you to create instance models of the whole system or instance models of any of the subsystems. You do so by choosing the appropriate system implementation as the root of your instance.

You use a tool like OSATE to generate the instance model. The generation is accomplished by first instantiating the elements that comprise a top-level system implementation and recursively instantiating their subcomponents and their subcomponents, and so on (e.g., systems to processes to threads). The resulting system instance model explicitly represents all of the requisite runtime aspects of the elements that comprise that system (e.g., binding of each thread to a processor, assigning required runtime properties to each element of the system).

2.2.2 AADL Textual and Graphical Representation

So far, we have focused on textual AADL. The well-defined semantics of AADL concepts is an important advantage of the AADL, especially for quantitative system architecture analysis. However, there are standard graphical representations that correspond to the textual representations of the elements of the AADL. These are defined in the SAE AADL standard [AS5506A]. Figure 2-2 shows a textual representation for a thread and its corresponding graphical representation. The graphical icon for a thread type is a dotted parallelogram as shown for the thread *data_processing*. Data ports are represented as small triangles that adorn a type icon, as shown for the data ports *raw_speed* and *speed_out*. The value of the *Period* of execution for a thread is contained inside an elliptical adornment at the top of the thread icon.

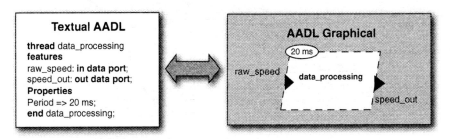

Figure 2-2: *AADL Graphical and Textual Representations*

In developing an AADL model of a system, either or both of these representations may be used. The preferred form is a matter of personal choice and is dependent upon the capabilities of the modeling tool you use. For example, you can create the initial structure of an embedded system architecture graphically and refine the model in textual AADL or rely exclusively upon the graphical interface for the creation and editing of AADL models and their corresponding textual representations. System instances are created and stored as system instance specifications (models) that are represented in XML. Analysis and code generation tools can operate on system instance models.

Throughout this book, we use the Open Source AADL Tool Environment (OSATE) as an example environment in our presentation of the AADL and its application in model-based engineering practices [OSATE][2]. The capabilities described are not unique to the OSATE environment and the use of AADL does not require OSATE. As shown in Figures 2-3 and 2-4, OSATE supports graphical as well as textual representations of AADL and provides editing capabilities for each of them[3]. For example, you can create a component implementation declaration for a system graphically (see Figure 2-3) by selecting modeling elements from a palette and identifying component types from component libraries (packages). You then fill in properties for the components

---

2. The Open Source AADL Tool Environment (OSATE) can be downloaded from www.aadl.info or the public AADL Wiki https://wiki.sei.cmu.edu/aadl.

3. There may be subtle differences in the implementation of the graphical representations in various tools (e.g., the label of *<data>* on a connection rather than the name of the connection).

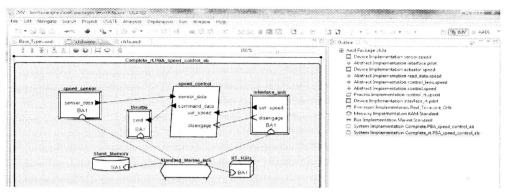

Figure 2-3: *OSATE Graphical Representation*

through a forms-based AADL property view. In other words, working with the graphical editor does not require detailed knowledge of the syntax of the AADL textual language in order to create initial models. You can then switch to the textual view (see Figure 2-4) and refine the AADL model textually when appropriate.

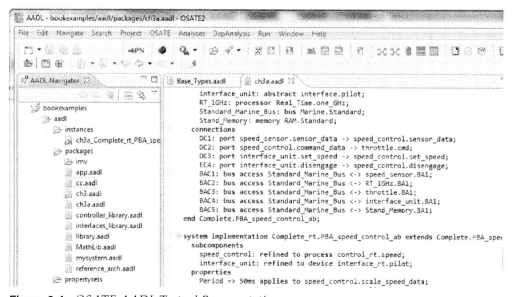

Figure 2-4: *OSATE AADL Textual Representation*

2.2.3 Analyzing Models

The well-defined semantics of AADL concepts allow various forms of quantitative analysis to be applied to AADL instance models. To create a system instance model for analysis, you can instantiate a partial or complete application system together with an execution platform declarative model. In this instance model, you can assign resource budgets in terms of CPU cycles and memory requirements to application subsystems and resource capacities to your execution platform. Given this data, you can analyze various bindings of application components to the execution platform and ensure that budgets do not exceed the platform's capacity. You can also add flow specifications to individual subsystem components and end-to-end flows to the application system. Based on these flow specifications, flow analyses such as an end-to-end response time analysis can be performed. All of these analyses can be completed without a fully detailed system model. At the completion of an architectural development, a complete system instance model can be analyzed in extensive detail to assess compliance with required functional and quality levels.

Early in the system development process, it is desirable to have partially specified but analyzable system models. For example, you may want to represent an application system as a collection of interacting subsystems without providing details of their implementation (e.g., where subsystems are modeled only as abstract or system components). Many of these analyses can be conducted on partial declarative models. For example, you can check that all ports are connected, that there is data consistency between the ports involved in a connection, and that the resource budgets are within the specified capacities of the system.

Chapter 3

Modeling and Analysis with the AADL: The Basics

In this chapter, we illustrate the development of basic AADL models and present general guidance on the use of some of the AADL's core capabilities. With this, we hope to provide a basic understanding of architectural modeling and analysis and start you on your way in applying the AADL to more complex software-dependent systems.

While reading the first part of this chapter, you may want to use an AADL development tool to create the specifications and conduct the analyses described. OSATE supports all of the modeling and analyses discussed in this chapter.

3.1 Developing a Simple Model

In this section, we present a step-by-step development and analysis of an AADL model. Specifically, we model a control system that provides a single dimension of speed control and demonstrate some of the analyses that can be conducted on this architectural model. The speed control functionality is part of a powerboat autopilot (PBA) system that is detailed in Appendix A. While specialized to a powerboat, this model exemplifies the use of the AADL for similar control applications such as aeronautical, automotive, or land vehicle speed control systems.

The approach we use is introductory, demonstrating the use of some of the core elements and capabilities of the AADL. We do not include many of the broader engineering capabilities of the language. For example, we do not address packages, prototypes, or component extensions in developing this simple model. These are discussed later in this chapter. Instead, we proceed through the generation of a basic declarative model and its instance and show a scheduling analysis of the system instance. During your reading of this section, you may want to reference Part II for details on specific AADL elements or analyses used in the example.

Initially we create a high-level system representation using AADL system, process, and device components. Building on this initial representation, we detail the runtime composition of all of the elements; allocate software to hardware resources; and assign values to properties of elements to a level that is required for analysis and for the creation of an instance of the system. In these steps, we assume that requirements are sufficiently detailed to provide a sound basis for the architectural design decisions and trade-offs illustrated in the example. In addition, while we reference specific architectural development and design approaches that put the various steps into a broader context, we do not advocate one approach over another.

3.1.1 Defining Components for a Model

A first step is to define the components that comprise the system and place their specification in packages. The process of defining and capturing components is similar to identifying objects in an object-oriented methodology. It is important to realize that components may include abstract encapsulations of functionality as well as representations of tangible things in the system and its environment. The definition of components is generally an iterative and incremental process, in that the first set of components may not represent a complete set and some components may need to be modified, decomposed, or merged with others.

First, we review the description of the speed controller for the PBA and define a simplified speed control model. In this model, we include a pilot interface unit for input of relevant PBA information, a speed sensor that sends speed data to the PBA, the PBA controller, and a throttle actuator that responds to PBA commands.

For each of the components identified, we develop type definitions, specifically defining the component's name, runtime category, and interfaces. Since we are initially developing a high-level (conceptual) model, we limit the component categories to system, process, and device.

The initial set of components is shown in Table 3-1, where both the AADL text and corresponding graphical representations are included. For this example, the textual specifications of all of the components required for the model are contained in a single package and no references to classifiers outside the package are required. Thus, a package name is not needed when referencing classifiers. For the graphical representations, the implementation relationship is shown explicitly. Note that the icon for an implementation has a bold border when compared to the border of its corresponding type icon.

The speed sensor, pilot interface, and throttle actuator are modeled as devices and the PBA control functions are represented as a process. We use the devices category for components that are interfaces to the external environment that we do not expect to decompose extensively (e.g., a device can only have a bus as a subcomponent).

Devices in AADL can represent abstractions of complex components that may contain an embedded processor and software. With a device component, you represent only those characteristics necessary for analysis and an unambiguous representation of a component. For example, in modeling a handheld GPS receiver, we may only be interested in the fact that position data is available at a communication port. The fact that the GPS receiver has an embedded processor, memory, touch screen user interface, and associated software is not required for analysis or modeling of the system. Alternatively, a device can represent relatively simple external components, such as a speed sensor, whose only output is a series of pulses whose frequency is proportional to the speed being sensed. If you require a complex interface to the external environment, you can use a system component. In this case, you can detail its composition and as needed include an uncomplicated device subcomponent to represent the interface to the environment.

The use of a process component for the control functions reflects the decision that the core control processing of the PBA is to be implemented in software. The software runtime components will be contained within an implementation of this process type. The implementation declarations in Table 3-1 do not include any details. As

Table 3-1: *Component Type and Implementations for the Speed Control Example*

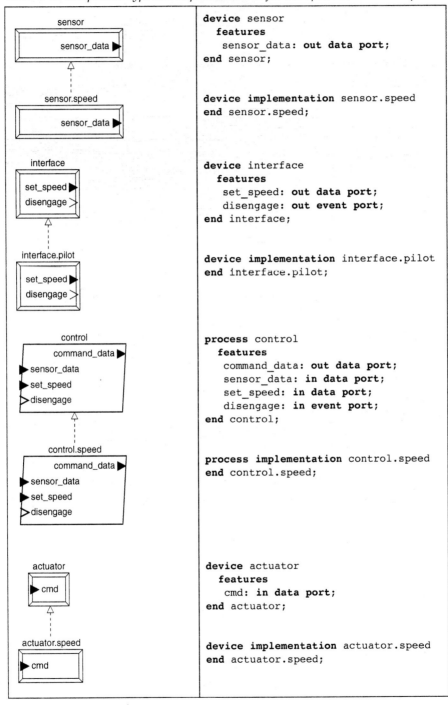

```
device sensor
  features
    sensor_data: out data port;
end sensor;
```

```
device implementation sensor.speed
end sensor.speed;
```

```
device interface
  features
    set_speed: out data port;
    disengage: out event port;
end interface;
```

```
device implementation interface.pilot
end interface.pilot;
```

```
process control
  features
    command_data: out data port;
    sensor_data: in data port;
    set_speed: in data port;
    disengage: in event port;
end control;
```

```
process implementation control.speed
end control.speed;
```

```
device actuator
  features
    cmd: in data port;
end actuator;
```

```
device implementation actuator.speed
end actuator.speed;
```

the design progresses, we will add to these declarations (i.e., adding subcomponents and properties as appropriate).

The interfaces for the PBA components are port features declared within a component type and are reflected in each implementation of that type. For example, the type *sensor* outputs a value of the speed via a data port *sensor_data*. The pilot interface type *interface* provides a value for the set speed via a data port *set_speed* and generates a signal to disengage the speed control via an event port *disengage*.

Notice that we have used explicit as well as abbreviated naming for the ports and other elements of the model (e.g., *command_data* and *cmd* for the command data at input and output ports). The specificity of names is up to you, provided they comply with AADL naming constraints for identifiers (e.g., the initial character cannot be a numeral). Note that naming is case insensitive and *Control* is the same name as *control*.

In the PBA example, we have chosen to assign specific runtime component categories to each of the components (e.g., the speed sensor is a device). However, in real-world development as a design matures, the definition of these components may change (e.g., a component that computes the PBA speed control laws may initially be represented as a system and later modified to a process or thread). Using the approach we outline here, these changes are done manually within the AADL model (i.e., changing a system declaration to a process category declaration). An alternative approach is to use the generic abstract component category (i.e., not defining a specific runtime essence). Then later in the development, converting this abstract category into a specific runtime category employing the AADL **extends** capability (e.g., converting an abstract component to a thread). We have chosen to use the former approach to simplify the presentation and focus on decisions and issues related to representations of the system as concrete runtime components. A discussion of the use of the abstract component category is provided in Section 3.5.

For each of the component types we define a single implementation. These declarations are partial, in that we omit substantial details needed for a complete specification of the architecture. For example, we do not define the type of data that is associated with the ports. We will address these omissions as required in later steps. However, we can conduct a number of analyses for our simple example without including many of these details.

3.1.2 Developing a Top-Level Model

In the next step, we integrate the individual component implementations into a system by declaring subcomponents instances and their connections. We do this by defining an enclosing system type and implementation as shown in Listing 3-1, where we define a system type *Complete* and its implementation *Complete.PBA_speed_control*. There is nothing special about our choice of naming for this enclosing system. Another naming scheme, such as a type of *PBA* and an implementation of *PBA.speed*, would work as well.

Within the implementation, we declare four subcomponents. The three device subcomponents represent the speed sensor, throttle, and the pilot interface unit. The process subcomponent *speed_control* represents the software that provides the speed control for the PBA. Notice that there are no external interfaces for the system type *Complete*. All of the interactions among the system's subcomponents are internal to the implementation *Complete.PBA_speed_control*, with the devices that comprise the system providing the interfaces to the external environment (e.g., sensors determining speed information from the vehicle).

Within the implementation, we define connections for each of the ports of the subcomponents. For example, connection *DC2* is the data connection between the *command_data* port on the process *speed_control* and the *cmd* data port on the device *throttle*. Each of the connections is labeled in the graphical representation shown in Listing 3-1 by the nature of the connection.[1] For example, connection *EC4* between the event port *disengage* on the *interface_unit* device and the event port *disengage* on the *speed_control* process is labeled as <<*Event*>>. It is our choice to match most of the port names. It is not required that connected ports have the same name. However, they must have matching data classifiers if specified (they are omitted in this initial representation).

Listing 3-1: *Subcomponents of the Complete PBA System*

```
system Complete
end Complete;

system implementation Complete.PBA_speed_control
   subcomponents
      speed_sensor: device sensor.speed;
```

---

1. The detailed graphical representation of the implementation *Complete.PBA_speed_control* is taken from the OSATE environment.

```
    throttle: device actuator.speed;
    speed_control: process control.speed;
    interface_unit: device interface.pilot;
  connections
    DC1: port speed_sensor.sensor_data ->
         speed_control.sensor_data;
    DC2: port speed_control.command_data -> throttle.cmd;
    DC3: port interface_unit.sct_speed ->
         speed_control.set_speed;
    EC4: port interface_unit.disengage ->
         speed_control.disengage;
end Complete.PBA_speed_control;
```

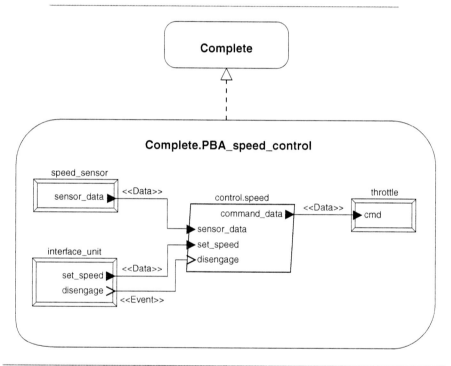

Depending upon your development environment the graphical portrayals may differ from those shown in Listing 3-1. For example, within OSATE you cannot display the containment explicitly. Rather, the internal structure of an implementation is presented in a separate diagram that can be accessed hierarchically through the graphical icon representing the implementation *Complete.PBA_speed_control*.

3.1.3 Detailing the Control Software

At this point, we begin to detail the composition of the process *speed_control*. This involves decisions relating to partitioning the functionality and responsibilities required of the PBA system to provide speed control. Since we have treated the speed control as an autonomous capability, we have implicitly assumed that there are no interactions between the directional or other elements of the PBA and the speed control system. This may not be the case in advanced control systems. In addition, for the purposes of this example, we partition the functions of the speed control process into two subcomponents. The first is a thread that receives input from speed sensor; scales and filters that data; and delivers the processed data to the second thread. The second is a thread that executes the PBA speed control laws and outputs commands to the throttle actuator. Again, this simplification may not be adequate for a realistic speed control system (e.g., the control laws may involve extensive computations that for efficiency must be separated into multiple threads or may involve complex mode switches that are triggered by various speed or directional conditions)[2].

Since the interfaces for the two threads are different, we define a type and implementation for each, as shown in Listing 3-2. We have used property associations to assign execution characteristics to the threads. Each is a periodic thread (assigned using the *Dispatch Protocol* property association) with a period of 50ms (assigned using the *Period* property association).

The assignment of periodic execution and the values for the period of the threads reflect design decisions. Generally, these are based upon the input of application domain and/or control engineers. The assignments we use here are not necessarily optimal but are chosen to provide specific values to enable analysis of system performance. They do not reflect the values for any specific control system.

Listing 3-2: *PBA Control Threads Declarations*

```
thread read_data
  features
    sensor_data: in data port;
    proc_data: out data port;
```

---

2. In the next section we will demonstrate the addition of operational modes.

```
  properties
    Dispatch_Protocol => Periodic;
    Period => 50 ms;
end read_data;

thread implementation read_data.speed
end read_data.speed;

thread control_laws
  features
    proc_data: in data port;
    cmd: out data port;
    disengage: in event port;
    set_speed: in data port;
  properties
    Dispatch_Protocol => Periodic;
    Period => 50 ms;
end Control_laws;

thread implementation control_laws.speed
end control_laws.speed;
```

We detail the declaration of the process implementation *control.speed* that is presented in Table 3-1 to include the two thread subcomponents and their interactions (**connections**), as shown in Listing 3-3. There are five connections declared. Four of these connect ports on the boundary of the process with ports on the threads (i.e., *DC1, DC3, DC4,* and *EC1*). The fifth connects the out data port *proc_data* on the thread *scale_speed_data* to the in data port *proc_data* on the thread *speed_control_laws*.

Listing 3-3: *The Process Implementation control.speed*

```
process implementation control.speed
  subcomponents
    scale_speed_data: thread read_data.speed;
    speed_control_laws: thread control_laws.speed;
  connections
    DC1: port sensor_data -> scale_speed_data.sensor_data;
    DC2: port scale_speed_data.proc_data ->
      speed_control_laws.proc_data;
    DC3: port speed_control_laws.cmd -> command_data;
    EC1: port disengage -> speed_control_laws.disengage;
    DC4: port set_speed -> speed_control_laws.set_speed;
end control.speed;
```

continues

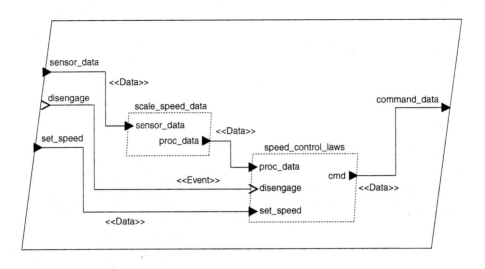

3.1.4 Adding Hardware Components

At this point, we have defined the software components of the speed control system. We now define the execution hardware required to support the software. In modeling the hardware and binding the control software to that hardware, we can analyze the execution timing and scheduling aspects of the system.

In Listing 3-4, we define a processor, memory, and bus. The processor will execute the PBA control code (threads) and the memory will store the executable code (process) for the system. In addition, we have declared that the processor type *Real_Time* and the memory type *RAM* require access to an instance of the bus implementation *Marine.Standard*. This bus will provide the physical pathway for the system. We will add properties to these declarations later in the modeling process.

Listing 3-4: *Execution Platform Declarations*

```
processor Real_Time
   features
      BA1: requires bus access Marine.Standard;
end Real_Time;
processor implementation Real_Time.one_GHz
end Real_Time.one_GHz;


memory RAM
   features
      BA1: requires bus access Marine.Standard;
end RAM;
```

```
memory implementation RAM.Standard
end RAM.Standard;

bus Marine
end Marine;

bus implementation Marine.Standard
end Marine.Standard;
```

3.1.5 Declaring Physical Connections

To continue the integration of the system, we add instances of the required execution platform components into the system implementation *Complete.PBA_speed_control* by declaring subcomponents for the implementation. In addition, we declare that these components are attached to the bus. This is done by connecting the **requires** interfaces on the processor and memory components to the bus component.

Since the PBA control software executing on the processor must receive data from the sensors and pilot interface unit as well as send commands to the throttle actuator, we declare that these sensing and actuator devices are connected to the bus as well. To do this, we add **requires bus access** declarations in the type declarations for these three devices and connect them to the bus. The updated declarations for the three devices are shown in Listing 3-5 and the graphical representation of the system with the declaration of the physical (**bus access**) connections is shown in Listing 3-6.

Listing 3-5: *Updated Device Declarations*

```
device interface
  features
    set_speed: out data port;
    disengage: out event port;
    BA1: requires bus access Marine.Standard;
end interface;

device sensor
  features
    sensor_data: out data port;
    BA1: requires bus access Marine.Standard;
end sensor;
```

continues

```
device actuator
  features
    cmd: in data port;
    BA1: requires bus access Marine.Standard;
end actuator;
```

In Listing 3-6, we have defined a processor *RT_1GHz*[3], bus *Standard_ Marine_Bus*, and memory *Stand_Memory* as subcomponents. In addition, we have declared the connections for the bus *Standard_Marine_Bus* to the **requires** bus access features of each of the physical components (e.g., from *Standard_Marine_Bus* to *RT_GHz.BA1* and to the **requires** bus access feature on the processor *RT_1GHz.BA1*). The **requires access** features and the bus access connections are shown in the graphical representation in the lower portion of Listing 3-6.

Listing 3-6: *Integrated Software and Hardware System*

```
system implementation Complete.PBA_speed_control
  subcomponents
    speed_sensor: device sensor.speed;
    throttle: device actuator.speed;
    speed_control: process control.speed;
    interface_unit: device interface.pilot;
    RT_1GHz: processor Real_Time.one_GHz;
    Standard_Marine_Bus: bus Marine.Standard;
    Stand_Memory: memory RAM.Standard;
  connections
    DC1: port speed_sensor.sensor_data ->
        speed_control.sensor_data;
    DC2: port speed_control.command_data -> throttle.cmd;
    DC3: port interface_unit.set_speed ->
        speed_control.set_speed;
    EC4: port interface_unit.disengage ->
        speed_control.disengage;
    BAC1: bus access Standard_Marine_Bus <-> speed_sensor.BA1;
    BAC2: bus access Standard_Marine_Bus <-> RT_1GHz.BA1;
    BAC3: bus access Standard_Marine_Bus <-> throttle.BA1;
    BAC4: bus access Standard_Marine_Bus <-> interface_unit.BA1;
    BAC5: bus access Standard_Marine_Bus <-> Stand_Memory.BA1;
end Complete.PBA_speed_control;
```

---

3. Since it is not the first character in the name, the numeric 1 can be used within the processor subcomponent name *RT_1GHz*. However, an implementation name *Real_ Time.1_GHz* is not legal, since the numeric is the first character in the implementation identifier.

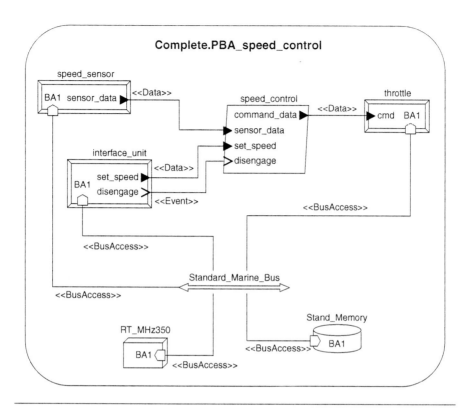

3.1.6 Binding Software to Hardware

In addition to specifying the physical connections, we bind software components to the appropriate physical hardware using contained property associations, as shown in Listing 3-7. These property associations are added to the system implementation declaration *Complete. PBA_speed_control.* The first two declarations allow the threads *speed_control_laws* and *scale_speed_data* to be bound to the processor *rt_mhz500.* The reference part of the property association identifies the specific processor instance *rt_mhz500* and the **applies to** identifies the specific thread in the hierarchy (e.g., **applies to** *speed_control.scale_speed_data* identifies the thread *scale_speed_data* that is located in the process *speed_control*). In this notation, a period separates the elements in the hierarchy.

We could have specified a specific binding using the *Actual_Processor_Binding* property. However, the *Allowed_Processor_Binding* property permits scheduling tools to assign the threads to processors. For example, the resourced allocation and scheduling analysis plug-in that is available in the OSATE environment[4] binds threads to processors taking into consideration the threads' period, deadline, and execution time; processor(s) speed and scheduling policies; and the constraints imposed by the actual and allowed binding properties[5]. Specifically, if only allowed processor bindings are defined (i.e., the *Allowed_Processor_Binding* property), the plug-in schedules the thread onto processors and reports back the actual thread to processor bindings and the resulting processor utilizations. If actual processor bindings are defined (i.e., the *Actual_Processor_Binding* property) the plug-in reports processor utilization based upon those bindings; allocates threads to processors; and runs a scheduling analysis to determine whether the bindings are acceptable.

Generally, a scheduling analysis or scheduling tool does scheduling analysis such that given a set of threads and their binding to processors (allowed or actual bindings), requisite attributes of the threads and processors (e.g., period, worst case execution time, processor cycle time, etc.), and defined scheduling policy, it determines if the set of threads meets the system's timing requirements. Typical scheduling policies include round-robin (RR), shortest job first (SJF), earliest deadline first (EDF), and rate monotonic (RM). The specific information required by and output from scheduling analysis tools vary.

The OSATE resource allocation and scheduling analysis plug-in makes binding decisions and in that process runs a scheduling analysis determining whether the binding is acceptable. This can be based on earliest deadline first (EDF) and rate monotonic scheduling (RMS) for periodic threads. In addition, it can conduct a rate monotonic analysis for periodic threads. This is useful for control system applications where all tasks are periodic, such as the PBA speed control example. In cases where threads are already bound to processors (i.e., using the *Actual_Processor_Binding* property), the plug-in determines schedulability for that specific deployment configuration.

If priority is assigned by hand and rate monotonic scheduling is used, another OSATE plug-in (priority inversion checker) enables the determination of whether the system has potential priority inversion. More sophisticated schedulability analysis tools are available for analyzing AADL models. A listing of these is available at https://wiki.sei.cmu.edu/aadl.

---

4. The resource allocation and scheduling analysis OSATE plug-in combines a bin-packing algorithm with scheduling algorithms. The OSATE tool is available for download from www.aadl.info.

5. Some scheduling policies may require additional properties, such as explicit priority assignment. The scheduling tool in OSATE assumes all periodic tasks without shared logical resources; other scheduling tools, such as Cheddar, accommodate the full set of tasks in AADL including tasks with shared data components.

The third entry in Listing 3-7, binds the code and data within the process *speed_control* to the memory component *Standard_Memory*. We chose to use the actual rather than the allowed memory binding property, since there is only one memory component in the system and, while an additional processor might be added, we do not anticipate additional memory components to be added.

Listing 3-7: *Binding Property Associations*

```
properties
  Allowed_Processor_Binding => (reference(RT_1GHz))
    applies to speed_control.speed_control_laws;
  Allowed_Processor_Binding => (reference(RT_1GHz))
    applies to speed_control.scale_speed_data;
  Actual_Memory_Binding => (reference(Stand_Memory))
    applies to speed_control;
```

3.1.7 Conducting Scheduling Analyses

Having defined the threads and established their allowed bindings to processors, we can begin to assess processor loading and analyze the schedulability of the system.

Before we proceed with a scheduling analysis, we define the requisite execution characteristics for the threads as they relate to the capabilities of the processors to which they may be bound. We specify this information through properties of the threads and processors. In this case, there is only one processor with an execution speed of 1GHz, as shown in Listing 3-8. Both threads are declared as *Periodic* with a period of 50ms. The default value in the AADL standard for the *Deadline* is the value of the *Period*. This value can be overridden by assigning a value to the *Deadline* using a property association. The execution time of the *read_data* thread ranges from 1 millisecond (ms) to 2 milliseconds (ms), whereas the *control_laws* thread's execution time ranges from 3ms to 5ms (as assigned using the *Compute_Execution_Time* property associations). These execution times are relative to the processor *Real_Time. one_GHz* declared in the model[6].

---

6. If an AADL model has a single type of processor (i.e., only one processor speed) then the execution time is with respect to that processor. If there are multiple processors with different speeds, you can specify an execution time for each processor type (using in binding) or specify an execution time with respect to one of the other processors (the reference processor) using a scaling factor that is associated with each processor type. There is a *Reference_Processor* property and a *Scaling_Factor* property for this purpose.

Execution time estimates for the threads can be based upon timing measurements from prototype code or historical data for similar systems (e.g., systems with the same or comparable processors). By conducting the analysis early in the development process, you can get quantitative predictions of a system's performance. This information can be updated and re-evaluated as the design progresses. These early and continuing predictions can help to avoid last minute problems during code implementation and system integration (e.g., during testing when deadlines are not met because the processor loading exceeds the capability of the processor).

Listing 3-8: *Updated Declarations for Analysis*

```
thread read_data
  features
    sensor_data: in data port;
    proc_data: out data port;
  properties
    Dispatch_Protocol => Periodic;
    Compute_Execution_Time => 1 ms .. 2 ms;
    Period => 50 ms;
end read_data;

thread control_laws
  features
    proc_data: in data port;
    cmd: out data port;
    disengage: in event port;
    set_speed: in data port;
  properties
    Dispatch_Protocol => Periodic;
    Compute_Execution_Time => 3 ms .. 5 ms;
    Period => 50 ms;
end Control_laws;
```

At this point, we have defined a declarative model for a simple speed control system including all of the components, properties, and bindings to describing a deployment configuration. From the top-level system implementation of this declarative model you create a system instance model and analyze it with the OSATE scheduler and scheduling analysis plug-in.

In Figure 3-1, we show the results of that analysis. It shows that the two threads in the system only use 14% of the processor capabilities. The worst case execution time for the two threads is 7ms, which is 14% of their 50 millisecond period.

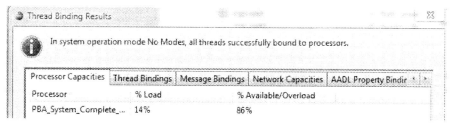

Figure 3-1: *Processor Capacities of the Speed Control System Instance*

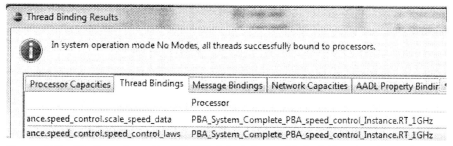

Figure 3-2: *Bindings from the OSATE Scheduler and Scheduling Analysis Plug-in*

Notice the information provided by the plug-in, including the actual binding for the threads determined by the plug-in (as shown in the cropped output of Figure 3-2).

3.1.8 Summary

At this point, we have developed a basic architectural model of the PBA speed control system. In so doing, we have demonstrated some of the core capabilities of the AADL. For this relatively simple model, we analyzed the execution environment and made predictions of schedulability of the system. In subsequent sections, we describe additional capabilities of the AADL and discuss alternative modeling approaches that can be applied in this simple example.

3.2 Representing Code Artifacts

Within a comprehensive AADL architectural specification, source code files and related information needed for specifying and developing the software within a system are documented using standard properties. These properties capture information for documenting architectural views such as code views [Hofmeister 00] and implementation views of allocation view types [Clements 10].

In this section, we document information relating to the PBA application software contained within the process *control.speed*. This excludes software that may be resident in the sensors, actuators, and interface devices as well as the operating system within the processor. First, we assume the application software has been written in a programming language, such as C or Java or in a modeling language, such as Simulink. We also assume that the software has been organized using the capabilities of the source language (e.g., by organizing Java classes and methods into packages with public and private elements). In this case, we can focus on specifying a mapping of the source files into the processes and threads of the application runtime architecture. Section 3.2.1 illustrates this mapping, which can be used to generate build scripts from the AADL model. Section 3.2.2 discusses how you can map identifier names used in AADL to identifier names that are acceptable in the source language. For larger systems, we may want to reflect not only the application runtime architecture in AADL, but also the modular source code structure. Section 3.2.3 illustrates how we utilize AADL packages for that purpose.

3.2.1 Documenting Source Code and Binary Files

A modified excerpt of the PBA specification is shown in Listing 3-9. This includes a properties section within the implementation *control. speed*, where the property association for the property *Source_Language* declares that the source code language for the implementation is C. This property is of type *Supported_Source_Languages*, which is defined in the property set *AADL_Project* and has the enumeration values (Ada95, C, Simulink_6_5 are some examples). Property types and constants in the *AADL_Project* property set can be tailored for specific projects. For example, languages such as Java can be added to the *Supported_Source_Languages* property type.

Using a property association for the newly defined property *Source_ Language,* the C language is declared as the programming language for all of the source code involved in the process *control.speed*. Two contained property associations for the property *Source_Text* identify the source and object code files for the threads *speed_control_laws* and *scale_ speed_data*. Two other contained property associations for the property

Source_Code_Size define the size of the compiled, linked, bound, and loaded code used in the final system.

In Listing 3-9, the data type *sampled_speed_data* is declared with a property association for the property *Source_Data_Size*. This property specifies the maximum size required for an instance of the data type. This data type is the classifier for the ports associated with the data that originates at the speed sensor.

Listing 3-9: *PBA Specification with Code Properties*

```
process implementation control.speed
  subcomponents
    scale_speed_data: thread read_data.speed;
    speed_control_laws: thread control_laws.speed;
  connections
    DC1: port sensor_data -> scale_speed_data.sensor_data;
    DC2: port scale_speed_data.proc_data ->
      speed_control_laws.proc_data;
    DC3: port speed_control_laws.cmd -> command_data;
    EC1: port disengage -> speed_control_laws.disengage;
    DC4: port set_speed -> speed_control_laws.set_speed;
  properties
    Source_Language => (C);
    Source_Text => ("ControlLaws.cc", "ControlLaws.obj")
      applies to speed_control_laws;
    Source_Text => ("ScaleData.cc", "ScaleData.obj")
      applies to scale_speed_data;
    Source_Code_Size => 4 KByte applies to scale_speed_data;
    Source_Code_Size => 10 KByte applies to speed_control_laws;
end control.speed;
```

3.2.2 Documenting Variable Names

A data port maps to a single variable in the application code. For example, the variable name for a data port can be specified using the *Source_Name* property. This is shown in Listing 3-10, for the in data port *set_speed* whose data classifier is the data type *set_speed_value*. The variable name for this port in the source code is *SetValue*.

We can use this mechanism to map data type and other component identifiers in an AADL model into the corresponding name in the source code. This is useful if the syntax of the source language allows characters in identifiers that are not allowed in AADL. We may also use this if we want to introduce more meaningful names in the AADL model for cryptic source code names.

Listing 3-10: *Example of Documenting Variable Names*

```
thread control_laws
  features
    proc_data: in data port;
    cmd: out data port;
    disengage: in event port;
    set_speed: in data port set_speed_value
      {Source_Name => "SetValue";};
  properties
    Dispatch_Protocol => Periodic;
    Period => 50 ms;
end control_laws;

data set_speed_value
end set_speed_value;
```

3.2.3 Modeling the Source Code Structure

Source code expressed in programming languages typically consists of data types and functions. They may take the form of subprograms and functions, classes and methods, or operations on objects. These source code elements are typically organized into libraries and modular packages. Some of the library or package content is considered public (i.e., it can be used by others), whereas other parts are considered private (i.e., can only be used locally). In the case of modeling languages such as Simulink, block libraries play a similar role.

Sometimes it is desirable to represent this modular source code structure in the AADL model. We can do so by making use of the AADL package concept. For example, we can model the functions making up a Math library by placing the subprogram type declarations representing the function signatures into an AADL package together with the subprogram group declaration representing the library itself, as shown in Listing 3-11.

We can place data component types that represent classes within the same source code package, into one AADL package. We can place the data component type and the subprogram types representing the operations on the source code data type in the same package. The methods of classes can be recorded as subprogram access features of the data component type (see Section 4.2.5). Any module hierarchy in the source code can be reflected in the AADL package naming hierarchy. For more on the use of AADL packages to organize component declarations into packages, see Section 4.4.1.

Listing 3-11: *Example of Modular Structure*

```
package MathLib
  public
  with Base_Types;
    subprogram group Math_Library
    features
      sqrt: provides subprogram access SquareRoot;
      log: provides subprogram access;
      pow: provides subprogram access;
    end Math_Library;

    subprogram SquareRoot
    features
      input: in parameter Base_Types::float;
      result: out parameter Base_Types::float;
    end SquareRoot;

end MathLib;
```

3.3 Modeling Dynamic Reconfigurations

Modes can be used to model various operational states and the dynamic reconfiguration of a system or component. In this section, we present the use of modes to represent the operation of the PBA speed control system. In this section, we develop another, slightly expanded model of the PBA speed control system.

3.3.1 Expanded PBA Model

We modify the PBA speed control model to include a *display_unit*. In addition, we add an out event port *control_on* to the *interface_unit*. Figure 3-3 shows the implementation of the expanded system including its subcomponents and their interconnections.

The type classifiers used in the expanded PBA model are shown in Listing 3-12. In this table, we define a process type *control_ex* that includes the additional features required for interfacing to the *display_ unit* and *interface_unit*. We could have declared this process type as an extension of the process type *control*, as shown in the comment. We also define the device type *interface_unit* as the device type *interface* with an additional port. Finally, we have added a new device type *display_unit*. The event port *control_on* is the trigger for a mode transition from *monitoring* to *controlling* and the event port *disengage* is the trigger for the reverse transition.

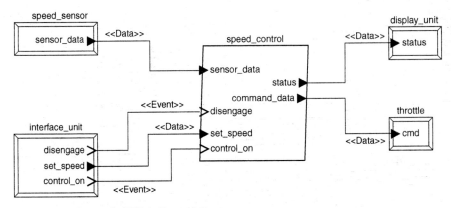

Figure 3-3: *Expanded PBA Control System*

Listing 3-12: *Type Classifiers for the Expanded PBA Model*

```
-- Type classifiers for the expanded PBA control system model --
process control_ex
  features
    sensor_data: in data port;
    command_data: out data port;
    status: out data port;       -- added port
    disengage: in event port;
    set_speed: in data port;
    control_on: in event port;   -- added port
  properties
    Period => 50 Ms;
end control_ex;

device interface_unit
  features
    disengage: out event port;
    set_speed: out data port;
    control_on: out event port;   -- added port
end interface_unit;

device display_unit            -- new device
  features
    status: in data port;
end display_unit;

thread monitor                 -- new thread
features
sensor_data: in data port;
status: out data port;
end monitor;
```

```
thread control_laws_ex
  features
    proc_data: in data port;
    set_speed: in data port;
    disengage: in event port;
    control_on: in event port;    -- added port
    status: out data port;        -- added port
    cmd: out data port;
end control_laws_ex;
```

Listing 3-12 also includes the new thread type *monitor*, and the modified thread type *control_laws* with an extra port, now called *control_laws_ex*. These are the thread types of the subcomponents of the process implementation *control_ex.speed*. A graphical representation of the subcomponents and connections for the process implementation *control_ex.speed* is shown in Figure 3-4. The process *speed_control*, shown in Figure 3-3, is an instance of *control_ex.speed*.

3.3.2 Specifying Modes

The textual specification for the implementation *control_ex.speed* is shown in Listing 3-13. In this implementation, two modes *monitoring* and *controlling* are declared in the **modes** section of the implementation. In the declarations for the subcomponents, the **in modes** declarations

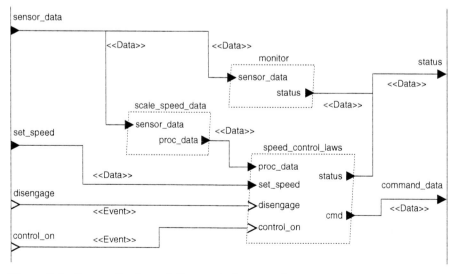

Figure 3-4: *Process Implementation control_ex.speed*

constrain the thread *monitor* to execute only in the *monitoring* mode and the threads *scale_speed_data* and *speed_control_laws* execute only in the *controlling* mode. Similarly, in the connection declarations are mode dependent such that the connections to the *monitor* thread are only active in the monitoring mode. The transitions between modes are triggered by the in event ports *control_on* and *disengage*. These are declared in the **modes** section of the implementation. If the modes are observable or are controlled from outside a component, then you may want to declare the modes in the component type.

A graphical representation for the mode transitions is shown in the lower portion of Listing 3-13. Modes are represented as dotted hexagons. The short arrow terminating at the monitoring mode denotes that the initial state is *monitoring*. The arrows connecting modes represent transitions. The input events are associated with the transitions that they trigger with a dotted line.

Listing 3-13: *Process Implementation of control_ex.speed with Modes*

```
process implementation control_ex.speed
  subcomponents
    scale_speed_data: thread read_data in modes (controlling);
    speed_control_laws: thread control_laws_ex
      in modes (controlling);
    monitor: thread monitor in modes (monitoring);
  connections
    DC1: port sensor_data -> scale_speed_data.sensor_data
      in modes (controlling);
    DC2: port scale_speed_data.proc_data ->
      speed_control_laws.proc_data in modes (controlling);
    DC3: port speed_control_laws.cmd -> command_data
      in modes (controlling);
    DC4: port set_speed -> speed_control_laws.set_speed
      in modes (controlling);
    DC5: port monitor.status -> status in modes (monitoring);
    DC6: port sensor_data -> monitor.sensor_data
      in modes (monitoring);
    DC8: port speed_control_laws.status -> status
      in modes (controlling);
    EC1: port disengage -> speed_control_laws.disengage
      in modes (controlling);
  modes
    monitoring: initial mode ;
    controlling: mode ;
    monitoring -[ control_on ]-> controlling;
    controlling -[ disengage ]-> monitoring;
  end control_ex.speed;
```

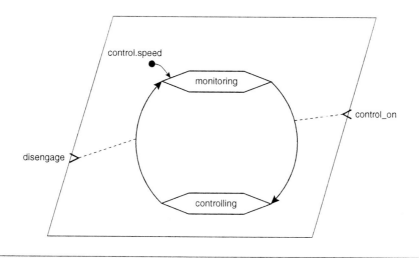

3.4 Modeling and Analyzing Abstract Flows

One of the important capabilities of the AADL is the ability to model and analyze flow paths through a system. For example within the PBA system, it is possible to analyze the time required for a signal to travel from the interface unit, through the control system, to the throttle actuator, and result in a throttle action.

3.4.1 Specifying a Flow Model

For this section, we add flow specifications to the expanded PBA speed control system shown in Figure 3-3. We investigate a flow path (the end-to-end flow) involving a change of speed via *set_speed* that extends from the pilot's interface unit to the throttle. In specifying the flow, we define each element of the flow (i.e., a flow source, flow path(s), flow sink), using *on_flow* as a common prefix for each flow specification name. In addition, we allocate transport latencies for each of these elements. Listing 3-14 presents an abbreviated version of the specification for the PBA speed control system that includes the requisite flow declarations.

The flow source is named *on_flow_src* that exits the interface unit through the port *set_speed*. The flow proceeds through the *speed_control* process via the flow path *on_flow_path* that enters through the port *set_speed* and exits through the port *command_data*. The flow sink occurs through the data port *cmd* in the *throttle* component. Note that a flow path can go from any kind of incoming port to a port of a different kind, for example from an event port to a data port.

Each flow specification is assigned a latency value. For example, the worst case time for the new speed to emerge from the interface unit after the pilot initiates the *set_point* request is 5ms. The worst case transit time through the *speed_control* process is 20ms and the time for the throttle to initiate an action is 8ms.[7]

Listing 3-14: *Flow Specifications for the Expanded PBA*

```
-- flow specifications are added to type declarations for this analysis ---

device interface_unit
  features
    set_speed: out data port;
    disengage: out event port;
    control_on: out event port;
  flows
  on_flow_src: flow source set_speed {latency => 5 ms .. 5 ms;};
end interface_unit;

process control_ex
  features
    sensor_data: in data port;
    command_data: out data port;
    status: out data port;
    disengage: in event port;
    set_speed: in data port;
    control_on: in event port;
  flows
  on_flow_path: flow path set_speed -> command_data
    {latency => 10 ms .. 20 ms;};
  properties
    Period => 50 Ms;
end control_ex;

device actuator
  features
    cmd: in data port;
    BA1: requires bus access Marine.Standard;
  flows
    on_flow_snk: flow sink cmd {latency => 8 ms .. 8 ms;};
end actuator;
```

7. These latency values are illustrative and do not reflect the performance of any particular device or speed control system.

3.4.2 Specifying an End-to-End Flow

The complete path, the end-to-end flow, for this example runs from the source component *interface_unit* through to the component *throttle*. This is declared in the system implementation *PBA.expanded*, as shown in Listing 3-15. The declaration originates at the source component and its source flow *interface_unit.on_flow_src* and the connection *EC4*. It continues through *speed_control.on_flow_path*, the connection *DC2*, and terminates at the sink *throttle.on_flow_snk*. In addition, we have specified a latency of 35ms for the flow. This value is drawn from the requirements for the system.

Listing 3-15: *An End-to-End Flow Declaration*

```
system implementation PBA.expanded
  subcomponents
    speed_sensor: device sensor.speed;
    throttle: device actuator.speed;
    speed_control: process control_ex.speed;
    interface_unit: device interface_unit;
    display_unit: device display_unit;
  connections
    DC1: port speed_sensor.sensor_data ->
      speed_control.sensor_data;
    DC2: port speed_control.command_data -> throttle.cmd;
    DC3: port interface_unit.set_speed ->
      speed_control.set_speed;
    EC4: port interface_unit.disengage ->
      speed_control.disengage;
    EC5: port interface_unit.control_on->
      speed_control.control_on;
    DC6: port speed_control.status -> display_unit.status;
  flows
    on_end_to_end: end to end flow
      interface_unit.on_flow_src -> EC5 ->
      speed_control.on_flow_path -> DC2 ->
      throttle.on_flow_snk {Latency => 35 ms .. 35 ms;};
end PBA.expanded;
```

3.4.3 Analyzing a Flow

At this point, we have defined a top-level end-to-end flow, assigned an expected (required) latency value to this flow, and defined latencies for each of the elements that comprise the flow. OSATE includes a flow latency analysis tool that automatically checks to see if end-to-end latency requirements are satisfied. For example, the tool will trace the path and total the latencies for the individual elements of the flow.

Figure 3-5: *Top Level End-to-End Flow Analysis Results*

This total is compared to the latency expected for an end-to-end flow. Figure 3-5 presents the results of this analysis for the PBA system we have specified. The total latency for the three elements of the flow at 33ms is less than the expected 35ms.

We could have manually determined, through a separate calculation, that the cumulative latency for the end-to-end flow would not violate the 35ms latency requirement. However, for very large systems, these calculations are difficult to do manually and it is difficult to ensure that latency values are correctly connected through the elements that comprise an architecture. The automated capabilities that can be included within an AADL tool facilitate easy calculation and re-calculation of these values. Moreover, having this data integral to the architecture provides a reliable way to manage the information and ensure consistency with updates to the data and the architecture.

3.5 Developing a Conceptual Model

It is possible to defer identifying the runtime nature of components until late in the development process. As noted earlier, you can do this by using system components and later manually changing the category in the relevant type, implementation, and subcomponent declarations for these components. In the next few sections, we present an alternative approach where you declare components as **abstract** and build an architectural component hierarchy. Then you use component extensions to create multi-view architectural representations. For example, in using the Siemens architecture approach [Hofmeister 00], you can include abstract components in a conceptual (runtime neutral) view and later extend these into runtime specific components, creating an execution view of the architecture.

3.5.1 Employing Abstract Components in a PBA Model

In declaring components for the PBA system, rather than using the device category for the pilot interface and the system category for the control components as we did in Table 3-1, we declare them as **abstract**. For this example, we assume there is a potential for decomposing the pilot interface into a complex interface unit. We could have made the sensor and actuator components abstract as well. However, to simplify the example and to demonstrate that you can mix abstract with runtime-specific categories, we maintain these components as devices. The declarations for this approach are shown in Table 3-2, where we have used the same partitioning and naming convention that is used in Table 3-1. Abstract components are represented graphically by dashed rectangles.

Table 3-2: *Abstract Component Declarations for the PBA*

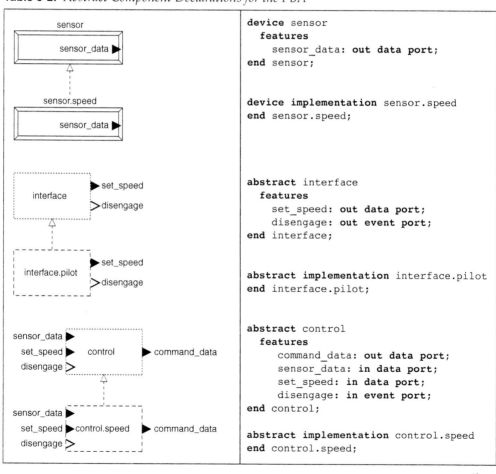

```
device sensor
  features
    sensor_data: out data port;
end sensor;


device implementation sensor.speed
end sensor.speed;




abstract interface
  features
    set_speed: out data port;
    disengage: out event port;
end interface;


abstract implementation interface.pilot
end interface.pilot;




abstract control
  features
    command_data: out data port;
    sensor_data: in data port;
    set_speed: in data port;
    disengage: in event port;
end control;


abstract implementation control.speed
end control.speed;
```

continues

Table 3-2: *Abstract Component Declarations for the PBA (continued)*

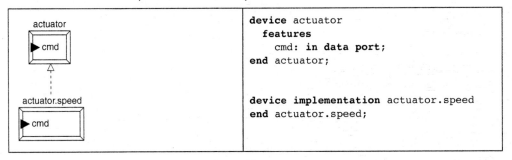

actuator — cmd
 	
actuator.speed — cmd
	```device actuator```
	 ``` features```
	 ``` cmd: in data port;```
	 ```end actuator;```
	
	 ```device implementation actuator.speed```
	 ```end actuator.speed;```

A complete system implementation using abstract components is shown in Listing 3-16. We have used an enclosing system, since we plan on instantiating it. However, we could have modeled the enclosing system as abstract as well, converting it to a system model later for instantiation. We have not included the hardware components or their relevant connections in this specification. We add these later in this discussion.

Listing 3-16: *Complete PBA System Using Abstract Components*

```
system Complete
end Complete;

system implementation Complete.PBA_speed_control_ab
  subcomponents
    speed_sensor: device sensor.speed;
    throttle: device actuator.speed;
    speed_control: abstract control.speed;
    interface_unit: abstract interface.pilot;
  connections
    DC1: port speed_sensor.sensor_data ->
      speed_control.sensor_data;
    DC2: port speed_control.command_data -> throttle.cmd;
    DC3: port interface_unit.set_speed ->
      speed_control.set_speed;
    EC4: port interface_unit.disengage ->
      speed_control.disengage;
end Complete.PBA_speed_control_ab;
```

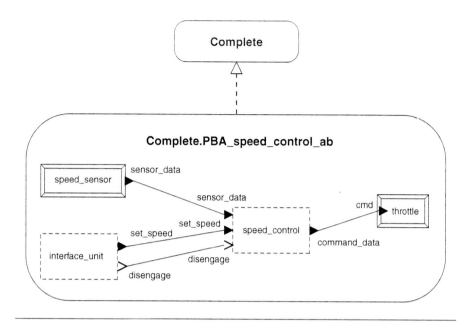

3.5.2 Detailing Abstract Implementations

In this section, we define the implementation *control.speed* for the *speed_control* subcomponent. This is shown in Listing 3-17. We detail this component by partitioning it into two subcomponents, as we did earlier in Listing 3-3. We declare the components *read_data* and *control* as abstract and include them as abstract subcomponents in the abstract implementation *control.speed*. In these declarations we have not included any property associations, specifically no runtime related properties. We will defer these until we generate the execution (runtime) representation. The interfaces and connections for data and control flow are included, since this information is nominally included in a conceptual (runtime neutral) representation. As before, no data types are defined for these interfaces[8].

---

8. Although we do not demonstrate this in the PBA example, the specific data types and data implementations can be added when the runtime categories are defined or even later in the process.

Listing 3-17: *Abstract Subcomponents for the control.speed Implementation*

```
abstract read_data
  features
    sensor_data: in data port;
    proc_data: out data port;
end read_data;

abstract implementation read_data.speed
end read_data.speed;

abstract control_laws
  features
    proc_data: in data port;
    cmd: out data port;
    disengage: in event port;
    set_speed: in data port;
end control_laws;

abstract implementation control_laws.speed
end control_laws.speed;

abstract control
  features
    command_data: out data port;
    sensor_data: in data port;
    set_speed: in data port;
    disengage: in event port;
end control;


abstract implementation control.speed
  subcomponents
    scale_speed_data: abstract read_data.speed;
    speed_control_laws: abstract control_laws.speed;
  connections
    DC1: port sensor_data -> scale_speed_data.sensor_data;
    DC2: port scale_speed_data.proc_data ->
                               speed_control_laws.proc_data;
    DC3: port speed_control_laws.cmd -> command_data;
    EC1: port disengage -> speed_control_laws.disengage;
    DC4: port set_speed -> speed_control_laws.set_speed;
end control.speed;
```

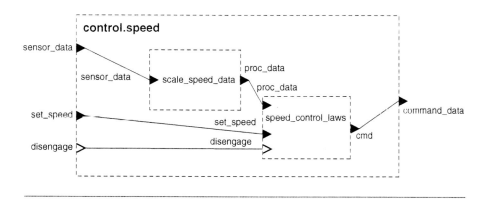

3.5.3 Transforming into a Runtime Representation

We transform abstract representations into runtime representations by extending implementations. In so doing, we change the category of an implementation and its corresponding type and transform the categories of the subcomponents that reference those classifiers. We start at the lowest level of the component hierarchy, extending implementations that have subcomponents. We progress upward until we reach the complete system. For this example, we use the same runtime categories as those in the previous section (i.e., those shown in Table 3-1).

First, we extend the implementation *control.speed*, since it is the lowest level abstract implementation with subcomponents.[9] This is shown in Listing 3-18, where a process type *control_rt* and a process implementation *control_rt.speed* are declared. The declaration of the type *control_rt* simply extends the type *control*, changing the category from abstract to process. There are no other refinements to the type. For the implementation *control_rt.speed*, the declaration changes the category of the implementation to process and refines (**refined to**) the category of both subcomponent to threads. Note that all of the characteristics (e.g., features, properties) of their ancestors are inherited by the components defined in an extension declaration (**extends**). Therefore, only modified elements of an implementation are included in an extension

---

9. We could have developed an abstract implementation for the pilot interface component *interface_pilot* that included subcomponents and a complex internal structure. In that case, we would refine it to a system or other runtime category.

declaration of that implementation. It is not necessary to extend the type or implementation declarations for the abstract implementations *read_data.speed* and *control_laws.speed*, since there are no subcomponents in either of these implementations.[10]

In changing a category, it is important that the features, subcomponents, modes, properties, etc. declared for an abstract component are consistent with the semantics of the new category. For example, an abstract component with a processor subcomponent cannot be extended into a thread component.

Next, we extend the enclosing system type *Complete* to create *Complete_rt* and its implementation *Complete.PBA_speed_control_ab* to create *Complete_rt.PBA_speed_control_ab*, as shown in Listing 3-18. In the declaration of *Complete_rt.PBA_speed_control_ab*, we also refine the subcomponents *speed_control* and *interface_unit*, changing their category from abstract to a runtime specific category. These choices parallel the categories of the simplified model developed earlier.

Listing 3-18: *Transforming the Generic PBA System into a Runtime Representation*

```
process control_rt extends control
end control_rt;

process implementation control_rt.speed extends control.speed
  subcomponents
    scale_speed_data: refined to thread read_data.speed;
    speed_control_laws: refined to thread control_laws.speed;
end control_rt.speed;


device interface_rt extends interface
end interface_rt;

device implementation interface_rt.pilot extends interface.pilot
end interface_rt.pilot;

system Complete_rt extends Complete
end Complete_rt;
```

---

10. In some cases, you may use abstract components that you decide should become processes, leaving them incomplete and detailing their implementation later (e.g., by adding subcomponents).

```
system implementation Complete_rt.PBA_speed_control_ab extends
Complete.PBA_speed_control_ab
  subcomponents
    speed_control: refined to process control_rt.speed;
    interface_unit: refined to device interface_rt.pilot;
end Complete_rt.PBA_speed_control_ab;
```

3.5.4 Adding Runtime Properties

At this point, we have refined the PBA model to include runtime components and subcomponents. However, we have not included runtime properties. For example, values for the timing properties required for a scheduling analysis are not assigned (e.g., the execution time for threads). We can do this in a number of ways. We could add local property associations to the individual abstract declarations, as shown in Listing 3-19 for the abstract types that are refined into threads. For properties that are declared as inheritable, we could modify components that are higher in the component hierarchy, relying on the inheritance of values to subcomponents (e.g., putting the values in the declarations for the abstract component type *control*).

Listing 3-19: *Modifying Declarations with Local Property Associations*

```
abstract read_data
  features
    sensor_data: in data port;
    proc_data: out data port;
  properties
    Dispatch_Protocol => Periodic;
    Compute_Execution_Time => 1 ms .. 2 ms;
    Period => 50 ms;
end read_data;

abstract control_laws
  features
    proc_data: in data port;
    cmd: out data port;
    disengage: in event port;
    set_speed: in data port;
  properties
    Dispatch_Protocol => Periodic;
    Compute_Execution_Time => 3 ms .. 5 ms;
    Period => 50 ms;
end Control_laws;
```

We could adopt a policy where we assign relevant properties by extending an abstract component (**extends**) and adding the property associations into the extension. This allows us to create multiple variants of the component parameterized with different property values. We can also parameterize individual subcomponents with different property values as part of a subcomponent refinement (**refined to**). An example of adorning the subcomponent refinements is shown in Listing 3-20.

Listing 3-20: *Property Associations Adorning Subcomponent Refinements*

```
process implementation control_rt.speed extends control.speed
  subcomponents
    scale_speed_data: refined to thread read_data.speed
      {Dispatch_Protocol => Periodic;
       Compute_Execution_Time => 1 ms .. 2 ms;
       Period => 50ms;};
    speed_control_laws: refined to thread control_laws.speed
      {Dispatch_Protocol => Periodic;
       Compute_Execution_Time => 3 ms .. 5 ms;
       Period => 50ms;};
end control_rt.speed;
```

Another approach to centralizing property associations is to include all property declarations for the extended system in the highest system implementation declaration or for a very large system in a limited number of system implementations. To do this we use contained property associations. This is useful when different instances of the same component need to have different property values. We effectively configure an instance of the model through properties and place this configuration information (property associations) in one place instead of modifying different parts of the model. An example is shown in Listing 3-21. We assign values to the *Period* and *Compute_Execution_Time* properties for the thread subcomponents using individual property associations. We use a single property association to apply the value *Periodic* to the *Dispatch_Protocol* property for both threads.

Listing 3-21: *Contained Property Associations within a System Implementation*

```
system implementation Complete_rt.PBA_speed_control_ab extends
                      Complete.PBA_speed_control_ab
  subcomponents
    speed_control: refined to process control_rt.speed;
    interface_unit: refined to device interface_rt.pilot;
```

```
  properties
    Period => 50ms applies to speed_control.scale_speed_data;
    Compute_Execution_Time => 1 ms .. 2 ms
      applies to speed_control.scale_speed_data;

    Period => 50ms applies to speed_control.speed_control_laws;
    Compute_Execution_Time => 3 ms .. 5 ms
      applies to speed_control.speed_control_laws;

    Dispatch_Protocol => Periodic
      applies to speed_control.scale_speed_data,
                 speed_control.speed_control_laws;
end Complete_rt.PBA_speed_control_ab;
```

3.5.5 Completing the Specification

In order to complete the specification for the PBA system to the level of the model we developed in the previous section, we need to include hardware components, their relevant interfaces, and their interconnections. For this purpose, we simply use the updated hardware component declarations as shown in Listing 3-4.

In addition, we need to add a bus access feature to the abstract component *interface.pilot*. A completed PBA speed control system implementation is shown in Listing 3-22. In the table, we have highlighted the portions of *Complete.PBA_speed_control_ab* that were modified in the extension to *Complete_rt.PBA_speed_control_ab*.

By adding the hardware subcomponents into the system implementation *Complete.PBA_speed_control_ab*, we have a system implementation *Complete_rt.PBA_speed_control_ab* that is comparable to the one we generated in the previous section. That is, with this representation, we can add binding properties and conduct a scheduling analysis as we did in Section 0.

Listing 3-22: *A Complete PBA System Implementation*

```
abstract interface
  features
    set_speed: out data port;
    disengage: out event port;
    BA1: requires bus access Marine.Standard;
end interface;

device sensor
  features
    sensor_data: out data port;
```
continues

```
      BA1: requires bus access Marine.Standard;
end sensor;

device actuator
  features
    cmd: in data port;
    BA1: requires bus access Marine.Standard;
end actuator;

system implementation Complete.PBA_speed_control_ab
  subcomponents
    speed_sensor: device sensor.speed;
    throttle: device actuator.speed;
    speed_control: abstract control.speed;
    interface_unit: abstract interface.pilot;
    RT_1GHz: processor Real_Time.one_GHz;
    Standard_Marine_Bus: bus Marine.Standard;
    Stand_Memory: memory RAM.Standard;
  connections
    DC1: port speed_sensor.sensor_data ->
      speed_control.sensor_data;
    DC2: port speed_control.command_data -> throttle.cmd;
    DC3: port interface_unit.set_speed ->
      speed_control.set_speed;
    EC4: port interface_unit.disengage ->
      speed_control.disengage;
    BAC1: bus access Standard_Marine_Bus <-> speed_sensor.BA1;
    BAC2: bus access Standard_Marine_Bus <-> RT_1GHz.BA1;
    BAC3: bus access Standard_Marine_Bus <-> throttle.BA1;
    BAC4: bus access Standard_Marine_Bus <-> interface_unit.BA1;
    BAC5: bus access Standard_Marine_Bus <-> Stand_Memory.BA1;
end Complete.PBA_speed_control_ab;

system implementation Complete_rt.PBA_speed_control_ab extends
Complete.PBA_speed_control_ab
  subcomponents
    speed_control: refined to process control_rt.speed;
    interface_unit: refined to device interface_rt.pilot;
  properties
    Period => 50ms applies to speed_control.scale_speed_data;
    Compute_Execution_Time => 1 ms .. 2 ms
      applies to speed_control.scale_speed_data;

    Period => 50ms applies to speed_control.speed_control_laws;
    Compute_Execution_Time => 3 ms .. 5 ms
      applies to speed_control.speed_control_laws;

    Dispatch_Protocol => Periodic
      applies to speed_control.scale_speed_data,
                 speed_control.speed_control_laws;
end Complete_rt.PBA_speed_control_ab;
```

3.6 Working with Component Patterns

As you use the AADL for multiple projects, you will find it convenient to reuse such things as data sensors, processors, buses, control software, and layered control architecture that have been successfully used in other projects. This is especially true if you are working in a product-line development environment.

In previous examples, we have seen how AADL can be used to define component templates (i.e., component descriptions that are completed and refined later through extension). In some cases, it is desirable to explicitly specify the placeholders (i.e., parameters) and what must be provided within a template. For example, we may have a template that is an abstract component defining a dual redundancy pattern. In that case, a user is expected to supply a single classifier that is used for both redundant instances in the pattern.

In this section, we discuss the use of parameterized component templates patterns. In so doing, we declare incomplete component types and implementations; explicitly specify what is needed to complete a pattern by declaring a prototype as a pattern parameter; and illustrate how such parameterized templates are used.

3.6.1 Component Libraries and Reference Architectures

With the AADL, it is possible to archive components and proven system solutions and reuse them through extension declarations. For this purpose, we suggest partitioning archival elements into two sets: a component library and reference architecture archive. The partitioning separates concerns such that individual, relatively simple elements are archived separately from elements involving a complex component hierarchy.

A component library is a collection of component types and component implementations with limited subcomponents that represent individual elements of a system architecture. These may be generic or runtime specific. For example, in a component library you may have a processor type *marine_certified* and a collection of implementations that have different processor speeds, manufacturers, and internal memory sizes. Similarly, you may have an abstract type *PID_controller* and its implementations that represent proportional-integral-derivative control with varying capabilities. The abstract components can be extended into runtime specific components such as a process or thread. For software components, this is the most flexible category for archiving in a library.

In your work, you may have identified a number of proven architecture solutions that have been useful. You can compile these solutions (reference architectures) into an archive that can be used in other projects. These reference architectures define common building blocks and reflect a common topology and are common throughout embedded systems development. Examples include layered control and triple modular redundant reference architectures that can be used for high dependability control avionics as well as space systems. Reference architectures can be defined at different levels of abstraction. Reference architectures can be defined using runtime-specific categories or abstract components and prototypes.

As a third approach to modeling the PBA speed control system, we use a generic component library, a reference architecture archive, prototypes, extensions, refinements, and multiple packages as demonstrations of reusing generic patterns for components and system architectures. We refine the generic components into runtime specific components in developing the PBA-specific architecture.

A library and archive can be developed without using prototypes (i.e., using only extensions and refinements). However, using prototypes makes explicit the elements (e.g., port and subcomponent classifiers) that are being refined.

3.6.2 Establishing a Component Library

Listing 3-23 shows an example generic component library that consists of two packages: *interfaces_library* and *controller_library*. In these packages, we define generic application components as **abstract** components. The packages are partitioned based upon a separation of concerns (e.g., the *interfaces_library* package has generic representations for sensors, actuators, and user interfaces). Another generic package could include only execution hardware with standard processors, memory, and bus components.

For this example, only the *generic_control* type has an implementation with subcomponents. In this implementation, prototypes are used in defining the subcomponents. Note that the property *Prototype_ Substitution_Rule* is assigned the value *Type_Extension*. This allows within refinements, the substitution of classifiers for prototypes that are of the same type or are an extension of the original type used for the prototype. Although most of the components in this library are abstract, runtime-specific categories can be used. For example, in the abstract type declaration for *generic_interface*, we define a data prototype *out_data* that is used to define the data type in the declaration of the out data port *output*.

Listing 3-23: *Generic Component Library*

```
--- generic component library ---

package interfaces_library
public

abstract generic_sensor
  features
  output: out data port;
end generic_sensor;

abstract generic_interface
  prototypes
    out_data: data;
  features
    output: out data port out_data;
    disengage: out event port;
end generic_interface;

abstract generic_actuator
  features
  input: in data port;
end generic_actuator;

end interfaces_library;

package controller_library
public

abstract generic_control
  features
    input: in data port;
    output: out data port;
    set_value: in data port;
    disengage: in event port;
end generic_control;

abstract implementation generic_control.partitioned
  prototypes
    rd: abstract generic_read_data;
    cl: abstract generic_control_laws;
  subcomponents
    r_data: abstract rd;
    c_laws: abstract cl;
```

continues

```
   connections
     DC1: port input -> r_data.input;
     DC2: port r_data.output -> c_laws.input;
     DC3: port c_laws.output-> output;
     EC1: port disengage -> c_laws.disengage;
     DC4: port set_value -> c_laws.set_value;
   properties
     Prototype_Substitution_Rule => Type_Extension;
 end generic_control.partitioned;

abstract generic_read_data
features
input: in data port;
output: out data port;
end generic_read_data;

abstract implementation generic_read_data.impl
end generic_read_data.impl;

abstract generic_control_laws
  features
     input: in data port;
     set_value: in data port;
     disengage: in event port;
     output: out data port;
end generic_control_laws;

abstract implementation generic_control_laws.impl
end generic_control_laws.impl;

end controller_library;
```

3.6.3 Defining a Reference Architecture

A sample reference architecture archive is shown in Listing 3-24, in which we have defined a generic speed control implementation *Complete.basic_speed_control_ref*. This implementation uses prototypes. The prototypes used here are abstract. However, prototypes can be runtime specific. In this reference architecture, we use the prototypes as classifier placeholders for the subcomponent classifiers of the implementation. For example, the prototype *ssg* represents the *generic_sensor* type that is defined in the package *interfaces_library*. This prototype is used in the declaration of the subcomponent *ss*. In using the reference architecture for the PBA, we refine the prototype into a specific runtime implementation. In this case, since we have assigned the value *Type_Extension* to the *Prototype_Substitution_Rule* property, we can substitute implementations of extensions of the component type declared in the prototype bindings.

Listing 3-24: *Reference Architectures*

```
--- reference architecture archive ---
package reference_arch
public
with interfaces_library, controller_library;
system Complete
end Complete;


system implementation Complete.basic_speed_control_ref
  prototypes
    ssg: abstract interfaces_library::generic_sensor;
    csg: abstract controller_library::generic_control;
    iug: abstract interfaces_library::generic_interface;
    acg: abstract interfaces_library::generic_actuator;
  subcomponents
    ss: abstract ssg;
    ac: abstract acg;
    cs: abstract csg;
    iu: abstract iug;
  connections
    DC1: port ss.output -> cs.input;
    DC2: port cs.output -> ac.input;
    DC3: port iu.output -> cs.set_value;
    EC4: port iu.disengage-> cs.disengage;
  properties
    Prototype_Substitution_Rule => Type_Extension;
end Complete.basic_speed_control_ref;

end reference_arch;
```

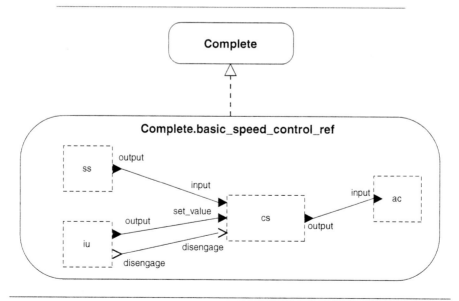

3.6.4 Utilizing a Reference Architecture

We use the reference architecture described in the previous section to define a PBA architecture. This is shown in Listing 3-25, where the first declaration extends the abstract type *Complete* found in the package *reference_arch*. In this extension, the system category is substituted for abstract. Similarly, the abstract implementation *Complete.basic_speed_control_ref* is extended creating the system *Complete.PBA_speed_control*. In this extension, the prototypes for the subcomponents are bound to an actual classifier using a prototype binding (e.g., *acg* => **device** *actuator.speed*). In our example, we have fixed the classifier to be a device called *actuator.speed*, which will be used in the subcomponent declaration that refers to the prototype.

In the second part of Listing 3-25, each of the type and some implementation classifiers used in the prototype refinements are extended from the component library. In these extensions, PBA specific refinements can be made. For example, the data classifier *speed_data* is added to the out port of the sensor in the *sensor* type and to the input of the process type *control*. In addition, property associations are added in the *control.speed* implementation. One is the period for the threads in the process *control.speed* and the other is a contained property association, assigning a compute execution time to the control laws thread *cl* within the process *control.speed*.

Listing 3-25: *Using a Reference Architecture*

```
package mysystem
public
with reference_arch, interfaces_library, controller_library;
system Complete extends reference_arch::Complete
end Complete;
system implementation Complete.PBA_speed_control
    extends reference_arch::Complete.basic_speed_control_ref
        (acg => device  actuator.speed,
         ssg => device sensor.speed,
         csg => process control.speed,
         iug => device interface.pilot )
end Complete.PBA_speed_control;

-- defining subcomponent substitutions ---

device sensor extends interfaces_library::generic_sensor
features
output: refined to out data port speed_data;
end sensor;
data speed_data
end speed_data;
```

```
device implementation sensor.speed
end sensor.speed;

device actuator extends interfaces_library::generic_actuator
features
input: refined to in data port cmd_data;
end actuator;

data cmd_data
end cmd_data;

device implementation actuator.speed
end actuator.speed;

device interface extends interfaces_library::generic_interface
end interface;

device implementation interface.pilot
end interface.pilot;

process control extends controller_library::generic_control
features
input: refined to in data port speed_data;
end control;

process implementation control.speed extends controller_
library::generic_control.partitioned
(
  rd => thread read_speed_data.impl,
  cl => thread speed_control_laws.impl )
properties
Period => 20 ms;
Compute_Execution_Time => 2ms..5ms applies to cl;
end control.speed;

thread read_speed_data
extends controller_library::generic_read_data
end read_speed_data;

thread implementation read_speed_data.impl extends controller_
library::generic_read_data.impl
end read_speed_data.impl;

thread speed_control_laws
  extends controller_library::generic_control_laws
end speed_control_laws;

thread implementation speed_control_laws.impl
    extends controller_library::generic_control_laws.impl
end speed_control_laws.impl;

end mysystem;
```

Chapter 4

Applying AADL Capabilities

In this chapter, we present additional design, modeling, and analysis considerations that provide a broad foundation for modeling and analyses with the AADL. We address issues relating to interactions among components, data modeling, system composition, and the organization of a system architectural model.

4.1 Specifying System Composition

In working with the AADL, you develop declarative models that describe individual software, hardware, and system components and their interactions and hierarchical organization. Within a declarative model, component type and implementation declarations are patterns (i.e., classifiers) for software and hardware component instances.

4.1.1 Component Hierarchy

A subcomponent declaration within a component implementation identifies an instance of the component pattern (i.e., a type or implementation) referenced in that declaration and designates that an instance of that pattern is contained within every instance of the implementation. These declarations within a declarative model describe the composition of a component as a hierarchy of component instances (i.e., subcomponent instances may contain subcomponents, etc.). For example consider Listing 4-1, the process implementation *control.speed* is a pattern that can be instantiated as a subcomponent of another implementation such as the system implementation *compete.PBA_ speed_control*. The subcomponent process instance *speed_control* itself contains an instance of the thread implementation classifier *control_ laws.basic*, resulting in the hierarchy where the system *complete.speed_ control* contains the process instance *speed_control* that contains the thread instances *scale_speed_data* and *speed_control_laws*.

Listing 4-1: *Component Hierarchy*

```
system implementation complete.PBA_speed_control
  subcomponents
    speed_sensor: device sensor.speed;
    throttle: device actuator.speed;
    speed_control: process control.speed;
    interface_unit: device interface.pilot;
  connections
    DC1: port speed_sensor.sensor_data ->
      speed_control.sensor_data;
    DC2: port speed_control.cmd -> throttle.cmd;
end complete.speed_control;

process implementation control.speed
  subcomponents
    scale_speed_data: thread read_data.speed;
    speed_control_laws: thread control_laws.speed;
  connections
    DC1: port sensor_data -> scale_speed_data.sensor_data;
    DC2: port scale_speed_data.proc_data -> speed_control_laws.
proc_data;
    DC3: port speed_control_laws.cmd -> command_data;
    EC1: port disengage -> speed_control_laws.disengage;
    DC4: port set_speed -> speed_control_laws.set_speed;
end control.speed;
```

4.1.2 Modeling Execution Platform Resources

Execution platform components and their representations encompass computer hardware and physical system components. Computer hardware consists of processors, memory, buses, and systems composed of them. The physical system components are represented by devices and logically interact with software as shown in Section 3.1. The same devices are physically connected to computer hardware via buses.

Execution platform components are abstractions of the computer hardware and may include associated software (e.g., software within devices and operating system software in processors). For example, in one situation you might represent a Linux-based computing resource simply as a processor component. Alternatively you may represent that resource as a system consisting of a core processor with Linux operating system software mapped onto that processor.

In defining a hardware configuration, you define hardware instances and the physical connectivity of those instances. That is, you declare processor, memory, and bus instances as subcomponents of a computing system. In the classifiers for these subcomponents, you declare that the processors and memory require access to the bus, the instances of these components in a system implementation, and the connectivity among these resources. For example, consider the system implementation *Execution_Hardware.PBA_redundant*, shown in the partial specification of Listing 4-2. This system implementation includes redundant processors and memory that are interconnected through a bus *Standard_Marine_Bus*. Each of the processor and memory subcomponents are connected to the bus, as shown graphically in the lower portion of Listing 4-2.

Listing 4-2: *Hardware Connections*

```
system implementation Execution_Hardware.PBA_redundant
  subcomponents
    RT_1GHz_primary: processor Real_Time.one_GHz;
    RT_1GHz_backup: processor Real_Time.one_GHz;
    Primary_Memory: memory RAM.Standard;
    Back_up_Memory: memory RAM.Standard;
    Standard_Marine_Bus: bus Marine.Standard;
  connections
    BAC1: bus access Standard_Marine_Bus <->
      RT_1GHz_primary.BA1;
    BAC2: bus access Standard_Marine_Bus <-> RT_1GHz_backup.BA1;
    BAC3: bus access Standard_Marine_Bus <-> Primary_Memory.BA1;
    BAC4: bus access Standard_Marine_Bus <-> Back_up_Memory.BA1;
end Execution_Hardware.PBA_redundant;

processor Real_Time
  features
    BA1: requires bus access Marine.Standard;
end Real_Time;
processor implementation Real_Time.one_GHz
end Real_Time.one_GHz;
```

continues

```
memory RAM
   features
      BA1: requires bus access Marine.Standard;
end RAM;

memory implementation RAM.Standard
end RAM.Standard;

bus Marine
end Marine;

bus implementation Marine.Standard
end Marine.Standard;
```

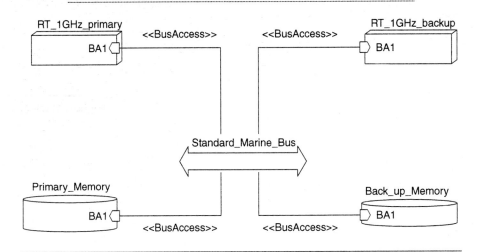

4.1.3 Execution Platform Support of Communication

In describing a system deployment, you must define the hardware that supports the execution of application components and the hardware that supports the logical connections between those application components. The hardware platform must include at least one processor to execute threads. A bus can support communications between two threads, each executing on a separate processor. To do this, you connect both processors to the same bus through a bus access connection and bind the connection between the threads to this bus.

Devices representing physical entities such as a digital camera may have a requires bus access feature to indicate that they communicate with a processor via a *usb* cable. A *usb* bus instance can then be used to establish a physical connection between the device and a processor.

A processor may provide access to one of its internal buses (bus sub-components) by declaring a provides bus access feature. Buses can be connected directly to other buses to represent complex inter-network communications, such as a PCI bus being connected to an Ethernet, which itself may be connected to a network backbone. Thus, a logical connection between application components can be bound to a sequence of buses or a sequence of buses with intervening processors.

4.1.4 System Hierarchy

Through the hierarchical modeling of its software and hardware components, you define the organization of a system. For a complete AADL system model, software components must be mapped onto execution platforms through binding relationships. These bindings define where code is executed and where data and executable code are stored within a system. For example, a thread must be bound to a processor for execution and a process must be bound to memory. Similarly, connections among components within a system must be bound to appropriate execution platform components (e.g., a simple connection is bound to a single bus or a connection within a complex distributed system is bound to a sequence of buses and intermediate processors and devices).

4.1.5 Creating a System Instance Model

You can create a system instance model by identifying the system implementation that is the top-level component of the system. Creating a system instance model is the process of creating component instances for the top-level system and recursively for each of its subcomponents using the component implementation declarations as blueprints. To this compositional hierarchy of components the instantiation process adds feature, connection, and flow instances. The resulting system instance represents the runtime architecture of an operational physical system.

You can create an instance model from a complete or a partially complete declarative AADL model. We can also create an instance model for a subsystem by itself by choosing its system implementation as the root of the instance model. We use these instance models to perform analysis, simulation, and auto generation of the embedded software system including its runtime executive.

4.1.6 Working with Connections in System Instance Models

A connection instance represents the actual flow of data and control between components of a system instance model. In the case of a fully specified system, a connection instance represents communication of events, data, or messages between two thread instances, a thread instance and a processor instance, a thread instance and a device instance, or two device instances. The data flow may be in either direction. In the AADL standard, connection instances in a fully specified system model are termed semantic connections.

A connection instance for a fully specified model is illustrated in Figure 4-1. In this figure, data is communicated between two threads in different processes. The data connection between the two threads is expressed by connection declarations that must follow the component hierarchy. In other words, there is a connection declaration from the original thread to its enclosing process, from that process to the second process, and from that process to the contained destination thread. Note that threads cannot arbitrarily communicate with other threads in the system. The enclosing process determines, through the ports in its type declaration and the connection declarations to those ports, which data from its threads should be passed on to threads in other processes.

In a system instance model, the sequence of data connection declarations from a thread to its enclosing process, to the second process, and to the thread contained in the second process results in a connection instance. If two threads are subcomponents within the same process or thread group, the connection instance is represented by a single connection declaration between those threads in the enclosing component implementation. While there may be a series of port-to-port

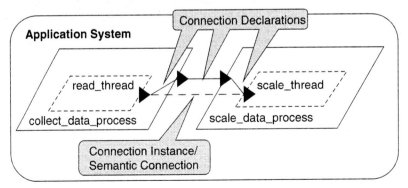

Figure 4-1: *A Semantic Connection Between Thread Instances*

connections involved in a data transfer (system instance connection) between two threads, data is transferred directly from the sending thread to the receiving thread. From an application source code perspective, the sending thread assigns a value to a variable/array and the receiving thread receives that value in a corresponding variable/array.

In the case of a partially specified system, the system instance model is expanded through the component hierarchy to the subcomponents for which no implementation detail is provided, regardless of their component category. In this case, connection instances may be between ports of system component instances or process component instances. According to the AADL standard, those connection instances are not semantic connections, but they are essential to certain analyses of partial system instance models.

As an example consider Figure 4-2, which illustrates a connection instance in a partial system instance model. In this model, the data collection process and the data scaling process have not been detailed. The data connection between the two processes results in a connection instance in the system instance model. This connection instance is not considered a semantic connection. However, the connection instance can be used in an analysis of this partially specified system such as a fault propagation analysis or flow analysis.

4.1.7 Working with System Instance Models

Early in the development process it is desirable to have partial system models and to instantiate them for analysis. For example, you may represent an application system as a collection of interacting subsystems

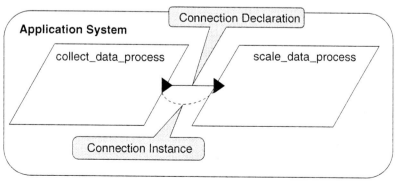

Figure 4-2: *A Connection Instance in a Partially Specified System Instance Model*

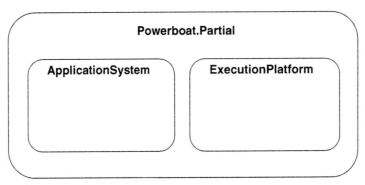

Figure 4-3: *Partial Powerboat Software and Execution Platform*

without providing details of their implementation. Subsystems are modeled as system components or process components. You can instantiate
this partial application system together with an execution platform
model into a partial system instance model as illustrated in Figure 4-3.
You can assign resource budgets in terms of CPU cycles and memory
requirements to the application subsystems and resource capacities to
the execution platform. Given this data, you can analyze various bindings of application components to the execution platform and ensure
that the budgets do not exceed the capacity. You can also add flow specifications to individual subsystem components and end-to-end flows to
the application system. Based on these flow specifications, flow analyses
such as an end-to-end response time analysis can be performed without
a fully detailed system model.[1]

4.2 Component Interactions

In this section, we discuss modeling three logical interactions between
application components

1. The directional exchange of control and data

2. Information exchange through shared data areas

---

1. For more information on analysis, see AADL publications and presentations at
www.aadl.info and the tool section of the public AADL Wiki at https://wiki.sei.cmu.
edu/aadl.

3. Service requests or function invocation including local calls, remote calls, and object-oriented method invocation

In addition, we discuss modeling interactions with the external environment and the physical infrastructure that supports these interactions.

4.2.1 Modeling Directional Exchange of Data and Control

The directional exchange of data and control is expressed using connections between application components (e.g., between threads) that involve event, data, and event data ports. Event port connections represent the asynchronous transfer of control. Data port connections represent the transfer of state data. Event data port connections represent the transfer of data as messages. In this section, we discuss some of the timing considerations associated with port communication between periodic threads focusing on issues relating to communication between periodic threads with differing execution frequencies.

Event and event data port connection transfers from a thread are immediate. This is the immediate transfer of control in the case of an event port and the immediate transfer of both control and data for an event data port. Event ports are used for event and alarm transmission and include the queuing of events at the port of the receiving thread. These events may trigger a dispatch[2] of a thread or trigger a mode transition. Message passing is represented using event data ports that map to message buffers or queues. The arrival of a message may trigger the dispatch of the receiving thread or device. The receiving thread may also sample the event or event data port periodically and process the entire contents of the queue. You can use this periodic sampling to model system health monitors that process alarms in process control system.

The transfer of data via data ports can be time coordinated. This is especially useful to ensure time-deterministic exchange and to minimize jitter (i.e., non-deterministic variations in the arrival time of data). The timing of data communication via data ports is expressed in terms of execution completion, deadline, and dispatch times and is dependent upon the type of components involved (i.e., thread, device, or processor) and the nature of their connections. For data port transfer out of threads, the data is ready for transfer at the completion of the thread. The type of data port connection establishes the timing of the delivery

---

2. Dispatch is used here to mean the initiation of execution.

of data to a receiving component. The data port connection types are sampled, immediate, or delayed.

The timing of sampling connections between components is independent of the execution of the sending component such that the value delivered to the receiving component is the value available at its execution, regardless of the execution timing of the sending component. In contrast, you can ensure a well-defined data exchange between periodic threads with simultaneous dispatch using delayed and immediate port communications.

For immediate connections between periodic threads, data transmission is initiated when the source thread completes and the value delivered to the receiving thread is the value produced by the sending thread at its completion. For an immediate connection to occur, the threads must share a common (simultaneous) dispatch.

For a delayed data port connection between periodic threads, the value from the sending thread is transmitted at its deadline and is available to the receiving thread at its next dispatch. For delayed port connections, the communicating threads do not need to share a common dispatch. In this case, the data available to a receiving thread is that value produced at the most recent deadline of the sending thread. If the deadline of the sending thread and the dispatch of the receiving thread occur simultaneously, the transmission occurs at that instant.

4.2.2 Modeling Shared Data Exchange

The local or a remote sharing of a data is modeled using data components and access connections that declare the logical and physical access to resources. For example, in the case of access to data resident in a memory unit, access to that memory must be declared. While shared data can be useful, in many cases, it is more efficient to use explicit data exchange (e.g., directional data transfer or messages) rather than shared data areas.

To establish a shared data area, you declare a data type and, if desired, a data implementation. You then define the specific instance of that data and declare the requisite accesses that define the shared access to the data instance. You define data instances as data subcomponents in threads, thread groups, processes, and systems. For example, you can declare a shared data component in a process that is used by several of the threads in the process. The threads indicate their need to access the data component through requires data access feature declarations. This is illustrated in Listing 4-3.

You can specify through the *Required_Access* property on the data access feature whether the thread intends to have read or write access to the data component. Given that several threads may access the same data component concurrently, a *Concurrency_Control_Protocol* property is available to specify the desired mechanism for ensuring mutually exclusive access. These are shown in Listing 4-3. The connection for *autoCruise.globalState* is shown as bi-directional, reflecting read-write access. However, for *receiveGPSSignal.globalState* the connection is directional, reflecting the write-only access to the data *currentState*.

Listing 4-3: *Shared Data Component*

```
process cruiseControl
end cruiseControl;

process implementation cruiseControl.simple
subcomponents
  currentState: data CruiseState
         { Concurrency_Control_Protocol => Semaphore;};
  receiveGPSSignal: thread GPSInterface;
  autoCruise: thread CruiseController;
connections
  da1: data access  receiveGPSSignal.globalState ->
     currentState;
  da2: data access currentState <-> autoCruise.globalState;
end cruiseControl.simple;

thread GPSInterface
features
  globalState: requires data access CruiseState
       { Access_Right => Write_Only;};
end GPSInterface;

thread CruiseController
features
  globalState: requires data access CruiseState
       { Access_Right => Read_Write;};
end CruiseController;
```

4.2.3 Modeling Local Service Requests or Function Invocation

Service requests or function invocations are modeled using calls to subprograms. When modeling calls to subprograms you can simply specify a reference to a subprogram classifier to identify the subprogram, or you can explicitly declare instances of subprograms and identify the instance in the call. When you specify the subprogram classifier, we assume that the linker/loader will determine how many copies of the subprogram are needed and which copy is invoked by the call. Typically, each process image has its own copy of subprogram code unless shared libraries are used.

Note that in an architecture model typically you would limit yourself to modeling (remote) subprogram calls across threads and processes. Modeling of subprogram calls within a thread represents detailed design, which is better captured using the Behavior Annex [BAnnex] or a detailed design model such as a Simulink model that you associate with the thread.

Listing 4-4 illustrates local subprogram calls within the thread *assess.basic*. The calls are shown to reference a subprogram classifier. The first call *do_review* identifies the subprogram *review.basic*. You define the signature of the subprogram by the subprogram type *review* with two parameters. You can add the subprogram implementation *review.basic* to model how this subprogram consists of calls to other subprograms. The second subprogram call *get_total* only refers to a subprogram type. This is sufficient if you do not need to model the internals of the subprogram as part of the architecture model, leaving the internal representation to a detailed design description. Similarly, you can leave off the parameters if the parameter passing details are not relevant to modeling the task and communication architecture of your application.

Listing 4-4: *Calls Within a Thread*

```
package app
public
thread assess
end assess;
--
thread implementation assess.basic
calls mainseq: {
do_review: subprogram library::review.basic;
get_total: subprogram library::total;
        };
end assess.basic;
end app;

package library
public
subprogram review
features
  report: in parameter;
  result: out parameter;
end review;
```

```
subprogram implementation review.basic
end review.basic;

subprogram total
end total;
end library;
```

Subprogram instances are modeled as a subcomponent in a thread or in an enclosing process. An example is shown in Listing 4-5, where the thread implementation *assess.basic* has a call to the subprogram instance *callreview* that is declared as subcomponent inside the thread implementation. The subprogram instance *callreview* is only callable within the thread, since the thread does not make the instance accessible from outside. It can be made accessible through a provides subprogram access feature. The second call refers to a requires subprogram access feature called *calltotal*. The process implementation *myapp.basic* contains an instance of the thread as subcomponent (*myapp*) and must provide the subprogram instance to the thread through the subprogram access connection *subcall*.

Listing 4-5: *Calls to Subprogram Instances Within a Thread*

```
thread assess
features  calltotal: requires subprogram access library::total;
end assess;
--
thread implementation assess.basic
subcomponents
callreview: subprogram review.basic;
calls main: {
do_review: subprogram callreview;
get_total: subprogram total;
      };
end assess.basic;

process myapp
end myapp;

process implementation myapp.basic
subcomponents
mytotal: subprogram library::total;
mythread: thread assess.basic;
connections
 subcall: subprogram access mytotal <-> mythread.calltotal;
end myapp.basic;
```

4.2.4 Modeling Remote Service Requests and Function Invocations

You can model a subprogram that can be called remotely as a provides subprogram access feature of a thread. This indicates that the subprogram can be called remotely by subprogram calls from other threads. In this case, the calling thread is suspended until the called thread has completed the execution of the subprogram and the call returns the results.

You can model a reference to a remote subprogram to be called in two ways. First, you can explicitly declare subprogram access connections to define the calling path to the specific instance called. This may be desirable if you are following a strict component-based modeling approach in which all interactions with other components including use of external subprograms should be specified as part of a component interface. Appropriate provides and requires subprogram access must be declared as a feature of the components involved and their containing components as appropriate. An example of a remote client-server call is shown in Listing 4-6, where the thread subcomponent *calling_thread* (an instance of the thread implementation *calling.impl*) calls the subprogram *service_it.impl*. The specific instance that is called is defined by the connection declarations from the requires subprogram access feature of the calling thread (i.e., from the *serveracc* feature of the client thread *calling_thread* in the process subcomponent *client_process*) through to the provides subprogram access feature of the server thread (i.e., to the *service* feature of the server thread *server_thread* in the process subcomponent *server_process*).

Second, you can model remote calls as a call binding to a remote service without having to clutter the model with subprogram access features and connections. You provide a value for the property *Actual_Subprogram_Call* that is associated with the call, as shown in the text insert and in the graphical representation of Listing 4-6. The property association is included in the properties section of the system implementation *client_server_sys.impl*. This association declares that the subprogram call *call_server* within the thread *calling_thread*, which is a subcomponent of the process *client_process*, is being made to the subprogram contained within the server process (*server_process*). You can place all remote subprogram call bindings in one place in the properties section of the top-level system implementation.

Listing 4-6: *Client-Server Subprogram Example*

```
system implementation client_server_sys.impl
subcomponents
client_process: process client_process.impl;
server_process: process server_process.impl;
connections
client_server: subprogram access client_process.serveracc ->
    server_process.service;

properties
Actual_Subprogram_Call =>
    reference (server_process.server_thread.service)
    applies to  client_process.calling_thread.call_server;
end client_server_sys.impl;
--
process client_process
serveracc: requires subprogram access service_it.impl;
end client_process;
--
process implementation client_process.impl
subcomponents
calling_thread: thread calling.impl;
connections
to_server: subprogram access calling_thread.serveracc ->
    serveracc;
end client_process.impl;
--
thread calling
features
serveracc: requires subprogram access service_it.impl;
end calling;
--
thread implementation calling.impl
calls   {
        call_server: subprogram service_it.impl;
      };
end calling.impl;
----
process server_process
features
service: provides subprogram access service_it.impl;
end server_process;
--
process implementation server_process.impl
subcomponents
server_thread: thread server_thread.impl;
connections
from_server: subprogram access server_thread.service -> service;
end server_process.impl;
--
```

continues

```
thread server_thread
features
service: provides subprogram access service_it.impl;
end server_thread;
--
thread implementation server_thread.impl
subcomponents
service: subprogram service_it.impl;
end server_thread.impl;
--
subprogram service_it
end service_it;

subprogram implementation service_it.impl
end service_it.impl;
```

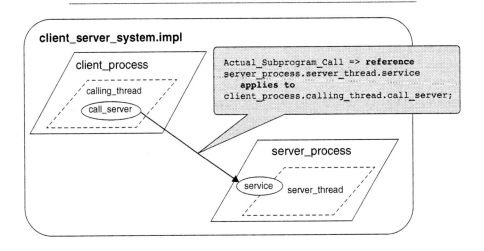

4.2.5 Modeling Object-Oriented Method Calls

You can represent calls to object methods using calls to subprograms contained within a data component. The data is made available to the subprogram using the name *this* for a **requires data access** feature to the data (by reference) or using a subprogram parameter to pass the data (by value) to the subprogram. Consider the pseudocode in Table 4-1 and a corresponding AADL model. For the class *ErrorLog*, the method *errorTotal* returns an integer value (e.g., the total number of errors currently in the log) and the reset method (e.g., sets the error count to zero). Calls to these methods are shown in the pseudocode for main.

The corresponding AADL model uses the **requires data access** (*this*) option. The specification defines an enclosing process implementation (a container for the calling thread and data object), a data subcomponent *ErrorData* (corresponding to an instance object of the class *ErrorData*), a thread subcomponent monitor (the thread that calls the methods in *ErrorData*), and the data access connection from *ErrorData* to the thread monitor. The implementation of the thread monitor includes a call sequence to the subprograms *errorTotal* and reset, reflecting the call sequence in main. The integer type return value for *errorTotal* is represented as an out parameter total for the subprogram. Each of the subprograms has a required data access feature *this*. This feature reflects the fact that a call to the subprogram requires access to the data instance of which they are a feature, in this case, the data subcomponent *ErrorData* that is an instance of the data component *ErrorLog*. Data access connections establish the call path from the individual subprogram calls to the requisite methods of the data instance *ErrorData*. For this example, these are *C1* in the process implementation *maintenance.control* (defines the path from the data instance *ErrorData* to the requires data access feature *log_access* of the thread monitor) and two connections within monitor from the feature *log_access* to the feature associated with each of the subprogram calls. These are shown graphically in the figure of Table 4-1.

The name this is used for the requires data access features of the subprograms to highlight the need for access to an instance. The corresponding AADL model defines an enclosing process implementation (a container for the calling thread and data object), a data subcomponent *ErrorData* (corresponding to an instance object of the class *ErrorData*), a thread subcomponent monitor (the thread that calls the methods in *ErrorData*), and the data access connection *this* to *ErrorData*.

A subprogram call can explicitly reference an instance of a subprogram by referring directly to a subprogram subcomponent or to a subprogram access feature. In this case, instances of subprograms must be declared explicitly as subcomponents. These subprogram instances may be shared between threads and other components. Threads and other components indicate that they require access to a subprogram or provide access to a subprogram through provides and requires subprogram access features. Subprogram access connections resolve the requires access to subcomponents by defining a path for that access. This supports a component-based development approach, in which it is desirable to completely specify all dependencies to other components, including subprogram call dependencies.

Table 4-1: *Methods Calls on an Object*

Object-Oriented Pseudocode	AADL Representation

<pre>class ErrorLog {
      int errorTotal () {
…
  }
      void reset() {
….
  }
…..</pre>
	<pre>process implementation maintenance.control
	subcomponents
	monitor: thread monitor.errors;
	ErrorData: data ErrorLog;
	connections
	C1: data access ErrorData <->
	                    monitor.log_access;
	end maintenance.control;</pre>

<pre>public static void main() {
…
ErrorLog stabilizer = new
ErrorLog();
int errors;
…
errors =
stabilizer.errorTotal();
stabilizer.reset();
...
  }</pre>
	<pre>--
	thread monitor
	features
	log_access: requires data access ErrorLog;
	end monitor;
	--
	thread implementation monitor.errors
	calls {
	errors: subprogram ErrorLog.errorTotal;
	reset_it: subprogram ErrorLog.reset;
	      };
	Connections
	  da1: data access log_access <->
	                    reset_it.this;
	  da2: data access log_access <->
	                    errors.this;
	end monitor.errors;
	--
	data ErrorLog
	features
	errorTotal: subprogram errorTotal;
	reset: subprogram reset;
	end ErrorLog;
	--
	subprogram errorTotal
	features
	  this: requires data access ErrorLog;
	  total: out parameter
	        BaseTypes::integer;
	end errorTotal;
	--
	subprogram reset
	features
	  this: requires data access ErrorLog;</pre>

Table 4-1: *Methods Calls on an Object (continued)*

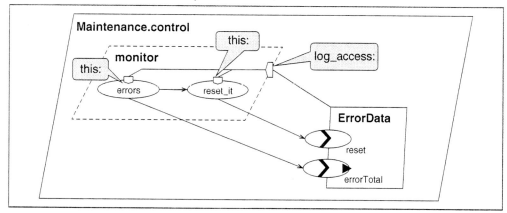

4.2.6 Modeling Subprogram Parameters

You can specify the signature of subprograms using parameter features and requiring data access features. Parameter features represent incoming and outgoing value parameters in source code. You can represent referenced parameters in source code as requires data access features.

   You can model the details of code sequences in the source code of threads with AADL. However, in doing so you are starting to represent design details rather than modeling at an architectural level. As an example, consider a subprogram that can receive and provide data through passing by value or passing by reference. In this case, the flow of data into and out of a subprogram can involve references to data (e.g., pointer values) or access to common data values (i.e., global or static data), rather than explicit data passing. These data reference mechanisms are described through data requires/provides data access declarations in an AADL model.

   For example, consider the annotated pseudocode and corresponding AADL textual representation in Table 4-2. In the pseudocode, examples of subprogram calls with data reference and the use of global data are shown. In the Passing by reference section of pseudocode, the function scale modifies data (referenced with the pointer *p1*) using the scale factor *v1*. In the second implementation of scale (the Global variable section of Table 4-2), a parameter data value (the scale factor) is passed and a common data element *raw_data* is scaled.

Table 4-2: *Examples of Passing by Reference and Global Data*

Pseudocode	AADL Representation
Passing by reference: .... scale (v1, p1) v1 is a real that is the scale factor. p1 is a pointer to a data set 'raw_data' that is to be scaled. ... processing that calls the subprogram: ... call scale (v1, p1);	**subprogram** scale **features** v1: **in parameter real;** p1: **requires data access** raw_data { Access_Right => Read_Only;}; **end** scale; -- **data** raw_data **end** raw_data; -- **data** real **end** real; -- **thread** processing **features** scalar: **in data port** real; p1: **requires data access** raw_data; **end** processing; --

Global variable:
 	
...
variable and processing
definitions:
....
real: raw_data;
...
scale(v1)
  {
    x := raw_data;
  }
...
processing that calls the
subprogram:
...
call scale(v1);
....
	thread implementation processing.impl
	calls {
	    scale_it: **subprogram** scale;
	    };
	connections
	VC1: **parameter** scalar -> scale_it.v1;
	PC1: **data access** p1 <-> scale_it.p1;
	end processing.impl;
	--
	process data_management
	features
	scalar: **in data port** real;
	end data_management;
	--
	process implementation data_management.impl
	subcomponents
	r_data: **data** raw_data;
	data_processing: **thread** processing.impl;
	connections
	C1: **data port** scalar -> data_processing.scalar;
	C2: **data access** r_data <-> data_processing.p1;
	end data_management.impl;

Within AADL, both of these options are represented with *v1* as a parameter, whereas the pointer *p1* and the common data *raw_data* are represented as a data access feature of the subprogram *scale*. The thread processing has a call to the subprogram *scale*. A corresponding AADL representation for the Global variable pseudocode explicitly shows the thread receiving the data value for *v1* through the in data port *scalar* and using that value in the subprogram call, as indicated by the parameter connection *VC1* in the thread. In contrast, the pointer reference to the data to be scaled is represented as a data access in the subprogram type declaration for *scale*. The explicit reference to *raw_data* in the subprogram *scale* is the requires statement in the thread type declaration. The AADL model allows an implementation using either option shown in pseudocode. You can reflect the fact that the subprogram scale only reads from *raw_data* by specifying an *Access_Right* property of *Read_Only* as part of the requires data access feature.

4.2.7 Interfacing to the External World

Interfaces outside of an application system often involve measuring domain specific parameters such as temperature, pressure, speed, and other physical parameters. In addition, for many application systems such as process and vehicle control systems there are outputs to physical devices. Within AADL, you can use a device component to represent these interfaces. The complexity and nature of a device depends upon how it and a system's external environment are included in the architecture. They can be a simple temperature sensor that outputs a value every 20 milliseconds or a complex GPS unit with satellite signal receivers and computing resources that provides the location of a vehicle to a jet engine controller.

Notice that the device has both physical bus connections to the hardware and logical port connections to the software. We recommend that you include the device as part of the application system, focusing on the logical connections. The device's physical connection to the execution platform is generally limited to a single bus connection, which follows the system hierarchy.

4.3 Modeling Data and Its Use

A variety of data types and data instances can be modeled using the data component category. Data classifiers can be used to define a data type (e.g., 16-bit integer values for temperature), model data embedded within another component (e.g., within a thread or as a record in another data component), or data that is shared among multiple components. We discussed the use of a data subcomponent to establish shared data in Section 4.2.2. In this section, we discuss the use of data classifiers to define data types and data subcomponents.

4.3.1 Defining a Simple Data Type

In defining a data type, you declare a data classifier (a type and optionally an implementation of that type) and reference that classifier. For example, you reference a data classifier in a data or event data port declaration and in so doing define the type of data that passes through the port (i.e., the type of the variable within the application code that represents the port). This is shown in Table 4-3, where the data types *comm_error* and *processor_error* are used to define the data type through the in data ports *comm_error_in* and *proc_error_in*. A data component type declaration may be as simple as introducing the name of the data type, as shown by the *comm_error* data type or more complicated as with *processor_error* that specifies the source code data variable name and size. When two data or event data ports are connected, their data types must match (i.e., they must refer to the same data classifier).

Data type declarations may include a property that specifies the size of an instance of this data type. For example, in Table 4-3 the data type *processor_error* is declared to have a size of 16 bits. This allows an analysis tool to determine the memory requirements for static data or the amount of data to be communicated through a port without having to specify the internal details of the data type.

Table 4-3: *Data Type Declarations*

Graphical Representation	AADL Textual Specification
comm_error	`data comm_error` `end comm_error;`

processor_error	`data processor_error`
	`properties`
	`Source_Name => "p_error";`
	`Source_Data_Size => 16 bits;`
	`end processor_error;`

comm_error_in
proc_error_in / error_handler
	`process error_handler`
	`features`
	`comm_error_in: in data port comm_error;`
	`proc_error_in: in data port processor_error;`
	`end error_handler;`

You can also associate a base type with the data type or the port that uses the data type. The base type is the basic machine representation of the data. For example, the property *Data_Model::Base_Type => (classifier(Base_Types::SignedInt16))*; could be a property declaration in the data type *speed_data_type* (see Section 4.3.3 for details on *Base_Type* and *Base_Types*).

Data component types represent data types in source code. For example, ports are accessible in the source code of the application through port variables. The programming language may allow identifiers for data types that are not legal AADL identifiers. In that case, you can declare the AADL data component type with a legal AADL identifier and specify the respective programming language identifier in the source code through the *Source_Name* property, as shown for the data type *processor_error* in Table 4-3.

4.3.2 Representing Variants of a Data Type

You can use data component implementations to represent variants of a data type. Multiple variants can be declared for the same data type. In its simplest form a data component implementation introduces the name of the variant and possibly specifies a variant-specific property. In Table 4-4, you can see two variants of the data type *temperature*, each with a different size (16 and 18 bits). These may be used as types as shown for the data port declaration *F_temp_in* of the process type *manage_temp*.

Table 4-4: *Data Type Variants*

Graphical Representation	AADL Textual Specification
| | ```
data temperature
end temperature;

data implementation temperature.celsius
properties
   Source_Data_Size => 16 bits;
end temperature.celsius;

data implementation temperature.fahrenheit
properties
   Source_Data_Size => 18 bits;
end temperature.fahrenheit;

process manage_temp
features
F_temp_in: in data port temperature.fahrenheit;
end manage_temp;
``` |
```

4.3.3 Detailing a Data Type

We would like to point out that the focus of AADL is as an architecture language rather than a data modeling language. In many cases, you may want to limit yourself to defining data types without providing details of their internal structure. That information may already have been captured in a data dictionary, a data modeling language, or in the source code. In that case, data type information relevant to the runtime architecture can be captured in properties.

The Data Modeling Annex standard [DAnnex] has introduced a standard set of properties and a predeclared package for that purpose. For example, you can use the property *Base_Type* to indicate the base representation of the data value. The annex provides a package called *Base_Types* that defines a collection of data component types to represent common base representations such as Boolean, integer, float, signed and unsigned integer and float of various bit sizes, and so on. The property *Data_Representation* has been introduced by the Data Modeling Annex for you to specify the data representation of the data type. The property *Dimension* allows you to specify the dimension(s) of data representations such as arrays. A subset of the Data Modeling Annex property set is shown in Listing 4-7. They are used in the specification of the data type *winds* as shown in Listing 4-8.

Listing 4-7: *Data Modeling Properties*

```
property set Data_Model is
Base_Type : list of classifier ( data )
    applies to ( data, feature );
Data_Representation : enumeration (Array, Boolean, Character,
    Enum, Float, Fixed, Integer, String, Struct, Union)
    applies to (data);
Dimension: list of aadlinteger applies to (data);
-- more property definitions
end Data_Model;
```

Listing 4-8: *Using Data Modeling Properties in a Data Type*

```
data winds
properties
 Data_Model::Base_Type => ( classifier(BaseTypes::Unsigned_16))
 Data_Model::Data_Representation => Array;
 Data_Model::Dimension => 10;
end winds;
```

Table 4-5: *Sample Data Component Declarations*

Graphical Representation	AADL Textual Specification

comm_error	```data comm_error```
	```properties```
	```Source_Data_Size => 16 bits;```
	```end comm_error;```

processor_error	```data processor_error```
	```properties```
	```Source_Data_Size => 16 bits;```
	```end processor_error;```

error_data	```data error_data``` ```end error_data;```

error_data.control
comm_dat
proc_dat
	```data implementation error_data.control```
	```subcomponents```
	```comm_dat: data comm_error;```
	```proc_dat: data processor_error;```
	```end error_data.control;```

If you do want to use AADL to model the internal structure of a data type, you can make use of the data component implementation declaration with subcomponents to represent the data structure of a data type explicitly. For example, data subcomponents in a data component implementation can represent the instance variables of a class or the fields of a record. In defining a data subcomponent, you declare a data classifier (a type and optionally an implementation of that type) and reference that classifier in a subcomponent declaration. An example is shown in Table 4-5, where the data implementation *error_data. control* has two data subcomponents.

4.4 Organizing a Design

A variety of capabilities within the AADL facilitates organizing models within a complete system architecture. These include packages, alternative implementations, and extension (i.e., inheritance) capabilities.

4.4.1 Using Packages

Packages are used to organize the component specifications of an AADL model. Packages have public and private sections. From outside a package, you can only reference component classifiers placed in the public section of the package. However, you can only reference component classifiers in those packages that are listed in a *with* clause within your (the referencing) package. From within a package, you can reference classifiers in the package's public section but you can only access classifiers in a private section from within the private section. These two capabilities allow you to modularize the AADL model and to control the visibility of the components.

You can use these capabilities of packages in a variety of ways from simply partitioning classifiers into separate packages based upon model or application specific groupings or based upon model independent characteristics. For example, you may reflect the source code organization of data types and functions in AADL packages as in Section 3.2.3. The runtime architecture of the application consists of a collection of thread, thread group, and process declarations, which can be placed in different packages to reflect that different teams or team members are responsible for their development. You might group all elements of the physical system being controlled by the embedded software into a collection of packages using AADL devices, buses, and systems. Similarly, you can place the components of your computing platform into a set of packages representing libraries of parts and platform configurations. Alternatively, you can group components into packages based upon the design, stratifying the system. For example, a package can be defined for a flight manager subsystem using constituent component subsystems, packages that contain generic (common) descriptions, or packages containing only data types (e.g., a data dictionary).

Listing 4-9 shows you an example of a simple system organized into four packages. Package *DataDictionary::avionics_data* contains data type definitions as data component types. We could have split this data dictionary into one for the OEM and a second one for the supplier, which would then require coordination of their content. These data types are used to define the data types of ports in the packages *OEM::avionics_subsystem* and *Supplier::avionics_sensor*. The intention to reference data component types such as within the type *Flight_Manager* is declared in the *with* statement of each package. The package

Supplier::avionics_sensor contains the specification for a GPS, whose port declaration actually references one of the data types.

The OEM defines the interfaces of the flight manager subsystem and the navigation sensor processing process in the package *OEM::avionics_subsystems*, also identifying the data type of the data to transfer through the ports. The OEM maintains the implementations separate from the interface specifications in the package *OEM::Avionics_ subsystem_implementations*. The *with* statement of that package indicates that the OEM will utilize specifications from the package containing the interface specifications and from the supplier package. The implementation declaration must identify the component type, which is declared in the package *OEM::avionics_subsystem*. Listing 4-9 shows you the use of **renames** to make this identifier available locally without having to qualify the reference to the package. The comment shows an alternative way of achieving the same by declaring an extension of the *Flight_Manager* type first. This second alternative is recommended if you want to refine the interface specification in the process. The comment lines (-- ......) indicate that other declarations may be there but are not shown.

Listing 4-9: *Example Design Organization Using Packages*

```
package DataDictionary::avionics_data
public
  data raw_data
  end raw_data;

  data processed_data
  end processed_data;
end DataDictionary::avionics_data;

package Supplier::avionics_sensors
public
with DataDictionary::avionics_data ;
  device GPS
  features
    output_data: out data port DataDictionary::avionics_data::raw_data;
  end  GPS;

  device implementation GPS.mil
  end GPS.mil;
end Supplier::avionics_sensors;

package OEM::avionics_subsystems
public
```

continues

```
with DataDictionary::avionics_data ;
  system Flight_Manager
  features
    output_data: out data port
        DataDictionary::avionics_data::processed_data;
  end Flight_Manager ;

  process NavigationSensorProcessing
  features
    input_data: in data port DataDictionary::avionics_data::raw_data;
  end NavigationSensorProcessing;
end OEM::avionics_subsystems ;

package OEM::avionics_subsystem_implementations
public
with OEM::avionics_subsystems, Supplier::avionics_sensors;
Flight_Manager renames system OEM::avionics_subsystems::Flight_Manager;
--
-- alternative way of providing local Flight_Manager name
-- system Flight_Manager extends
OEM::avionics_subsystems::Flight_Manager
-- end Flight_Manager;
  system implementation Flight_Manager.common
    subcomponents
      NSP :
process OEM::avionics_subsystems::NavigationSensorProcessing;
      GPS : device  avionics_sensors::GPS.mil;
-- .........
  end Flight_Manager.common;
end OEM::avionics_subsystem_implementations ;
```

4.4.2 Developing Alternative Implementations

You can use multiple system implementations of a single type to define several deployment configurations for a system. There is considerable flexibility in doing so. For example, you can define a process type *controller* and declare two implementations *controller.basic* and *controller.robust* as shown in Listing 4-10. In this case, the type establishes a common set of interfaces and each implementation involves an alternative control policy and different property values. For the implementation *controller.basic*, a simple control policy is used (i.e., *basic_control.impl*) for the *control_policy* subcomponent, whereas for *controller.robust* a more complicated control policy is used. The value for the *Period* is different in each of the implementations. These alternative configurations within implementations of a type can be more extensive than shown, involving differences in other aspects of an implementation such as connections, modes, call sequences, and prototypes.

Listing 4-10: *Alternative Implementations*

```
process controller
features
  input: in data port;
  output: out data port;
  error_out: out event port;
end controller;

process implementation controller.basic
subcomponents
  read_input: thread read_data.impl;
  control_policy: thread basic_control.impl;
properties
  Period => 50 ms;
end controller.basic;

process implementation controller.robust
subcomponents
  read_input: thread read_data.impl;
  control_policy: thread robust_control.impl;
properties
  Period => 20 ms;
end controller.robust;
```

4.4.3 Defining Multiple Extensions

In addition to alternative implementations, you can define multiple extensions to a basic type and implementation template and develop a family of alternative configurations. These declarations can extend others, modifying the underlying type and implementations and adding characteristics to them. Individual implementations can be extended multiple times and extensions themselves can be extended. Implementation and type extensions can be integrated to create an interrelated set of component types and implementations.

Listing 4-11 shows example type and implementation extensions with accompanying type extension declarations for the PBA system.[3] Relationships among the declarations are shown graphically following the textual AADL model. Many of the details of the declarations are omitted to simplify the discussion. The type extension *PBA_control_ extended* adds in data ports to support the depth and fish data input for the deluxe and prestige versions. The implementation extensions for the deluxe and prestige versions, add subcomponents to support the

---

3. Details of the PBA system are provided in Appendix A.

fish and depth input and to manage the redundancy of the prestige version. In addition to the modifications shown, appropriate connections, property associations, and so on would be required in a complete model of the system.

There are constraints on these extensions, in that an implementation can only extend another implementation if the extension shares a common type with its ancestor. This is the case in Listing 4-11, where *PBA_control_extended.presige* extends the implementation *PBA_control_extended.deluxe*. They have a type common ancestor component type *PBA_control*.

Listing 4-11: *Multiple Implementation Extensions*

```
system PBA_control
  features
     speed: in data port speed_data;
     s_speed: in data port set_speed;
     position: in data port position_data;
     set_direction: in data port direction_data;
     speed_cmd: out data port speed_cmd;
     dir_cmd: out data port direction_cmd;
     display_data: feature group display_data;
end PBA_control;

system PBA_control_extended extends PBA_control
features
  depth: in data port depth_data;
  finder: in data port fish_location_data;
end PBA_control_extended;

system implementation PBA_control.basic
end PBA_control.basic;

system implementation PBA_control_extended.deluxe extends
              PBA_control.basic
subcomponents
  data_manager_depth: process data_mananger.depth;
  data_manager_fish: process data_mananger.fish;
end PBA_control_extended.deluxe;

system implementation PBA_control_extended.prestige extends
              PBA_control_extended.deluxe
subcomponents
  redundancy_management: process redundancy_manage.dual;
end PBA_control_extended.prestige;
```

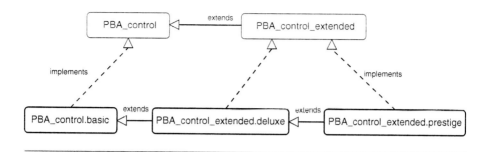

Part II

Elements of the AADL

The description of the elements of AADL provides a foundation for you to understand critical elements of the language, especially those of most value to a novice user. In addition, this part provides detailed reference material on the AADL that can be useful in reading Part I.

As an architecture language for embedded real-time system engineering, AADL was designed to be component based, capture all aspects of industrial-sized embedded systems, and support the validation through analysis and auto-generation of such systems from AADL models. Table II-1 organizes the elements of AADL according to these three perspectives: component modeling, embedded system modeling, and embedded system validation.

AADL supports component modeling with different categories of components with well-defined semantics to represent the application software, its runtime architecture in terms of communicating tasks, the computer platform of distributed processors and memory connected by buses, and the components of the physical target system. AADL supports embedded system modeling by capturing both static and dynamic aspects of an embedded system including component interactions, critical flows through the system, and deployment of software on

Table II-1: *Elements of the AADL*

Component Modeling	Components	• Component interfaces (*features*)
		• Component implementations
		  – Structure (*subcomponents*)
		  – Operational configurations (*modes*)
		• Generic and composite components (*abstract, system*)

	Application components	• Tasking (*thread, thread group, process*) • Software (*data, subprogram, subprogram group*)
	Execution platform components	• Computer hardware (*processor, memory, bus*) • Physical system components (*device*) • Partitions and protocols (*virtual processor, virtual bus*)
Embedded System Modeling	System architecture	• Static and dynamic architecture (*mode, instance*) • Component interaction (*port connections, access connections*) • Information flow (*flow specification, end to end flow*) • Software deployment (*bindings*)
	Model organization	• Component libraries (*packages*) • Parameterized templates (*prototypes*) • Refinements (*extends*) • Aliased references (*renames*)
Embedded System Validation	Model annotations	• Comments • Typed properties on components, features, connections, flows, modes • Mode-specific property values (*in modes*) • Sublanguage annotations (*annex library, annex subclause*)
	Language extensions	• User-defined properties (*property sets, property definition, property type*) • Sublanguage annexes (*error model, behavior*)
	Tooling	• Architecture model creation • Architecture analysis • System generation

hardware. Industrial scale embedded system modeling is supported by component libraries, parameterized templates, and model evolution through refinement. Model annotations make the architecture model the focus of system validation by supporting informal documentation as well as more quantified and formal specification. The extensibility of AADL through user-defined properties and specialized sublanguages facilitates the integration of different dimensions of operational quality attributes. This architecture-centric approach to system validation allows analytical models as well as runtime implementations to be auto-generated and validated from a single source.

The chapters of Part II follow the structure outlined in Table II-1. Chapter 5 introduces you to component interface specifications and blueprints of their implementation. Chapters 6 through 8 present the different categories of components. Chapters 9 through 11 focus on the static and dynamic architecture, component interactions, flows and deployment. Chapter 12 discusses different aspects of organizing large-scale system models. Chapters 13 and 14 describe mechanisms for annotating models with information relevant to validation through analysis. Chapter 15 provides an overview of tool-based capabilities for creating and validating system models captured in AADL.

Chapter 5

Defining AADL Components

A component is defined by its classifiers (i.e., a component type and a component implementation). A component type represents a spec sheet (i.e., the external interface of a component). It identifies the component category, defines all interaction points with other components, and externally observable characteristics. A component implementation represents a blueprint of its internal structure in terms of subcomponents (i.e., instances in terms of other component classifiers and connections between them and with the interfaces of the containing component implementation). The use of subcomponents results in a component hierarchy when an AADL model is instantiated. A component type can have multiple component implementations that act as variants of the component type. A component may not require a component implementation declaration if it is not composed of other components.

5.1 Component Names

Names for component types and implementations must consist of legal AADL identifiers and cannot be an AADL reserved word.[1] A legal

---

1. Legal AADL identifiers are described in Section 12.1 and AADL reserved words are listed in Appendix A.4.

AADL identifier is a single sequence (word) of letters or digits that begins with a letter and contains no spaces. An underscore can separate letters and digits in the sequence. The name of component types must be a single AADL identifier, whereas names for component implementations consist of two identifiers separated by a period (.). The first identifier is the name of the component type for the implementation, while the second identifier uniquely identifies the implementation within the type. For example, a component type may be named *sensor* and an implementation of that type may be *sensor.speed*.

You organize component classifiers into packages as component libraries. The classifier names must be unique within a package. You reference component classifiers within the same package by their component type identifier or by their period-separated component type and implementation identifier. Component classifiers in other packages are referenced by the package name followed by double colon (::) and the classifier name. When you declare a subcomponent, you may reference a component type if the referenced component does not have multiple component implementations.

5.2 Component Categories

Component categories establish the computational nature of a component. When you declare component types, component implementations, or subcomponent you specify the component category. Table 5-1 lists the component categories and organizes them into four sets. The application software, execution platform, and system sets provide the foundation for specifying the components that comprise a system architecture. The generic set consists of the abstract category that defines a runtime neutral component.

5.3 Declaring Component Types

A component type declaration defines a component's interface elements and externally observable attributes. The interface elements represent features (interaction points with other components), specifications of logical flows, and operational modes. Component properties represent observable attributes. Figure 5-1 shows the template for a component

Table 5-1: *Component Categories*

Application Software	`thread`	A schedulable unit of concurrent execution
	`thread group`	An abstraction for logically organizing threads, thread groups, and data components within a process
	`process`	Protected address space enforced at runtime
	`data`	Data in source code and application data types
	`subprogram`	Callable sequentially executable code that represents concepts such as call-return and calls-on methods
	`subprogram group`	An abstraction for organizing subprograms into libraries
Execution Platform (hardware)	`processor`	Schedules and executes threads and virtual processors
	`virtual processor`	Logical resource that is capable of scheduling and executing threads that must be bound to or be a subcomponent of one or more physical processors
	`memory`	Stores code and data
	`bus`	Interconnects processors, memory, and devices
	`virtual bus`	Represents a communication abstraction such as a virtual channel or communication protocol
	`device`	Represents sensors, actuators, or other components that interface with the external environment
Composite	`system`	Integrates software, hardware, and other system components into a distinct unit within an architecture
Generic	`abstract`	Defines a runtime neutral component

```
<component category> <type identifier>
 extends
 prototypes
 features
 flows
 modes | requires modes
 properties
 annex subclauses
 end <type identifier>;
```

Figure 5-1: *A Component Type Declaration*

type declaration. It may contain several optional sections marked by reserved words in the language. The reserved words of each section are bolded in Figure 5-1. The entries enclosed in angular brackets "<>" in Figure 5-1 are declaration specific, where a legal AADL component category replaces the entry <component category> and a legal AADL identifier that names the component type replaces the entry <type identifier>. A component category establishes the nature of a component (e.g., thread, processor, system). The vertical bar between the modes and requires modes statements indicates that if a modes section is declared, only one (i.e., modes or requires modes) can be included but not both.

Table 5-2 represents the sections that make up a type declaration and a brief description of their roles.

Table 5-2: *Sections of a Component Type Declaration*

Reserved Word	Description
extends	Enables one component type declaration to refine another type declaration inheriting its features and properties
prototypes	Allows named component types or implementations to be used as parameters within declarations
features	Specifies the interaction points for the component type (e.g., data transfer and access into and out of a component)
flows	Defines distinct abstract paths of information and control that pass into, out of, and through components of a system
modes	Specifies alternative configurations and transitions among them

Table 5-2: *Sections of a Component Type Declaration (continued)*

Reserved Word	Description
requires modes	Specifies the modes a component expects to inherit from its containing component
properties	Assigns values to various characteristics (properties) of a component and its related modeling elements.*
annex	Allows subclauses to be added in standardized or user-defined sublanguages, such as error modeling and behavioral specifications
end	Terminates the component declaration

* For example, a system type definition can assign a value to the property named Period. That value will apply to all subcomponents within all implementations of that type, unless explicitly overridden by another property value assignment.

The various sections of a type declaration shown in Figure 5-1 are optional and a component type declaration may have no entries, as shown in Table 5-3 for the thread type declaration *controller*. Typically, component types have a properties section and if they are not self-contained components, they have a features section. Sometimes you may define a component only partially and later refine it in a second component type declaration using the extends reserved word. The use of extends is illustrated in Section 12.3.1.

Rather than omitting a section in a component type, the entry "**none;**" can be used to explicitly indicate the omission. This is shown in Table 5-3 for the process type *PBA_speed_control* where the intent not to define properties or flows for this type is made explicit through the use of the entry "**none;**". Note that the entry (**none;**) can be placed on the same line as the reserved word that partitions the declaration. This practice can help to detect (and avoid) inadvertent omissions in a specification.

The graphical icons that correspond to the textual specifications are included in Table 5-3. As shown, the process icon is a solid parallelogram, whereas the thread icon is a dashed parallelogram. The declarations under **features** in the type declaration for *PBA_speed_control* define ports for the type. The next section describes ports and other features, including their graphical icons.

Table 5-3: *Sample Component Type Declarations*

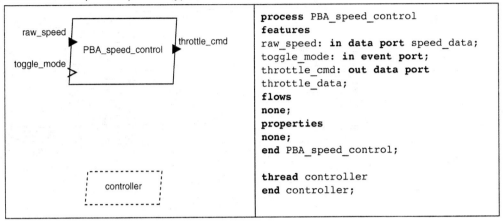

```
process PBA_speed_control
features
raw_speed: in data port speed_data;
toggle_mode: in event port;
throttle_cmd: out data port
throttle_data;
flows
none;
properties
none;
end PBA_speed_control;

thread controller
end controller;
```

5.4 Declaring a Component's External Interfaces

A component's external interfaces are the explicit entry and exit points for information and access for the component. You declare them within the features section of a component type declaration. Component features are divided into three categories: ports, access (to data, buses, and subprograms), and parameters. You may combine features into a feature group. A feature group itself can contain feature groups. Table 5-4 summarizes feature groups, component features, and their graphical representations.

Table 5-4: *A Summary of Component Features*

AADL Standard Graphical Notation	Feature	Description
◖	*feature group*	Feature groups represent groups of component features. Feature groups can contain feature groups. Feature groups can be used anywhere features can be used.
▶ > ▶▶	*data port* *event port* *event data port*	Ports enable the representation of the directional transfer of data and control (events) among components. Ports can be incoming and outgoing. The direction of transfer is indicated by the point.

Table 5-4: *A Summary of Component Features (continued)*

AADL Standard Graphical Notation	Feature	Description
⬠	*bus access* *data access*	Access features enable the specification of an application component's required and provided access to a common (shared) data component and a hardware component's access to a bus to represent physical connectivity.
◁❯▷	*subprogram access*	These features declare provided and required subprogram accessibility.
●❯❯	*subprogram group access*	These features declare provided and required subprogram library accessibility.
▶	*parameter*	Parameters enable the representation of the data associated with calls to subprograms.
● ●▷	*abstract feature*	An abstract feature can be refined into any of the concrete features. It may or may not include a direction. If unspecified, a direction is defined as part of the feature refinement.

The graphical representations for data port and parameters are the same. However, since parameters are features only of a subprogram and a data port cannot be a feature of a subprogram, when this icon adorns a subprogram icon it represents a parameter not a data port. In all other cases, the icon represents a data port.

The positioning of feature icons on the boundary of a component icon is shown in Table 5-5 using a system component. Feature icons can be placed anywhere on the component boundary. The icon pointing into or out of the component indicates the direction of incoming and outgoing ports.

The access icon pointing out of the component indicates required access, whereas for provided access the icon points into the component. In the case of feature groups, the ball of the icon always points out of the component, whereas the half circle is placed inside the component to indicate possible fan-out of individual elements of the feature group.

If a port represents both an incoming and outgoing port or parameter, the icon is placed back to back. Table 5-6 shows the resulting icons.

Table 5-5: *Feature Icons Positioning*

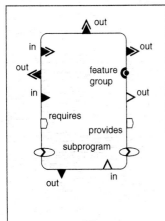	Graphically, a feature icon is positioned on the boundary of the component icon for which it is a feature. You can place them at any point along the boundary.
	Unidirectional ports, parameters, and access features are identified by the location and orientation of their icon on the boundary. For example, out port icons are located on the exterior face of a component's boundary point outward and in port icons are located on the interior face of a component's boundary and point inward. Similarly, requires access icons are located on the interior pointing inward and provides access icons are located on the exterior of a component boundary pointing outward.
	Subprogram access interfaces into/out of a component are oriented on the boundary with the direction of the arrowhead into/out of the component.
	Feature groups are located on the boundary with the half-circle portion of the icon on the interior of a component icon.

Table 5-6: *Bi-Directional Port Icons*

◆	In out data port and parameter	<>	In out event port	◀◆▶	In out event data port

A feature declaration includes a name, one or more defining reserved words (e.g., **in data port** or **requires data access**), relevant identifiers, and optional property associations for the feature. Sample feature declarations and related declarations are shown in Table 5-7. The process type declaration *PBA_speed_control* adds a requires data access feature to the type declaration provided in Table 5-3. This addition indicates the need for access to data of the type *error_log*. The subprogram type *read_it* includes a declaration of an input parameter to the subprogram with the name *size*. An implementation of this subprogram type *read_it* is used to represent a method call, as declared in the read subprogram feature declaration within the data type *error_log*. The second subprogram declaration defines a write method call on the data type *error_log*.

Table 5-7: *Sample Features Declarations*

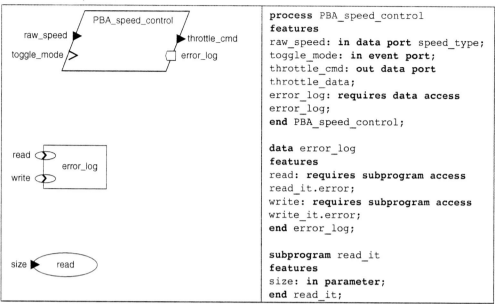

```
process PBA_speed_control
features
raw_speed: in data port speed_type;
toggle_mode: in event port;
throttle_cmd: out data port
throttle_data;
error_log: requires data access
error_log;
end PBA_speed_control;
```

```
data error_log
features
read: requires subprogram access
read_it.error;
write: requires subprogram access
write_it.error;
end error_log;
```

```
subprogram read_it
features
size: in parameter;
end read_it;
```

5.5 Declaring Component Implementations

A component implementation declaration defines a component's internal structure. Figure 5-2 shows the template for an implementation declaration. It may contain several optional sections marked by reserved words in the language. The entries enclosed in angular brackets "<>" in Figure 5-2 are declaration specific where a legal AADL category replaces the entry (component category), a component type identifier[2] replaces the entry (type identifier), and a legal AADL identifier that specifies the implementation replaces the entry (implementation identifier). The name of the implementation is the combination of type and implementation identifiers separated by a period.

Table 5-8 presents the sections that comprise a component implementation declaration and a brief description of their roles.

---

2. The component type identifier in a component implementation declaration must refer to a type declared in the same package or aliased from another package (see Section 12.2.3).

```
<component category> implementation <type identifier>.<implementation identifier>
    extends
    prototypes                                  implementation name
    subcomponents
    calls
    connections
    flows
    modes | requires modes
    properties
    annex subclauses
    end <type identifier>.<implementation identifier>;
```

Figure 5-2: *Structure of a Component Implementation Declaration*

Table 5-8: *Sections of a Component Implementation Declaration*

Reserved Word	Description
extends	Enables one component implementation to inherit the characteristics of another implementation declaration and all of its ancestors
prototypes	Allows named component types and implementations to be used as parameters
subcomponents	Specifies instances of components within a component implementation
calls	Defines sequences of calls to subprograms
connections	Establishes logical information exchanges and access between components
flows	Defines distinct abstract paths of information and control transfer
modes	Specifies alternative configurations and transitions among them
requires modes	Specifies the modes a component expects to inherit from its containing component
properties	Defines characteristics of a component and contained modeling elements
annex	Allows subclauses to be added in standardized or user-defined sublanguages such as error modeling and behavioral specifications
end	Terminates the component declaration

The various sections of an implementation declaration are optional. As with a type declaration, the entry "**none;**" can be used to explicitly indicate the omission of a section.

Listing 5-1 shows minimum example type and implementation declarations. However, a realistic component implementation declaration involves, at least, a set of properties to distinguish the implementation from other implementations of the same component type. If the component implementation represents a composite component (i.e., a component that is composed of other components) then the component implementation should contain declarations of subcomponents and connections.

Sometimes you may declare a component implementation partially and later refine it using the **extends** reserved word. This capability to have partial but syntactically legal specifications (i.e., where some sections and details are omitted) is helpful during incremental development and enables modeling and analysis early in the development process when many design details are not known.

Listing 5-1: *A Minimal Implementation Declaration*

```
thread controller
end controller;

thread implementation controller.speed
end   controller.speed;
```

Listing 5-2 shows a sample implementation declaration for the process type *PBA_speed_control* that was originally defined in Table 5-3. There are two thread subcomponents for the process. The type and implementation declarations for these threads are listed in the lower portion of Listing 5-2. The first three connections within the process implementation establish the data connections between ports on the boundary of the process and the thread subcomponents. The connection *C4* connects the output of the thread that reads the data to the input of the control law thread. Calls are not permitted within a process implementation (they need to be defined in the context of a thread) and are omitted in this example. In the thread type declaration *control_laws*, access to the data component state is declared as a feature. A subprogram feature is declared for the state data component and in the thread implementation *control_laws.speed*, a call to that subprogram is declared. The subprogram *log_state* is declared separately.

A graphical representation of the component implementation *PBA_speed_control* is shown in the lower portion of Listing 5-2. The process implementation icon has the thread subcomponents contained within it. The connections among the threads and between the threads and the ports on the implementation *PBA_speed_control.basic* are shown as solid lines between the features. Note that the graphical icon for the process implementation has a thicker border than the icon for the process type, as shown in Listing 5-2.

Listing 5-2: *Sample Component Implementation Declaration*

```
process PBA_speed_control
features
raw_speed: in data port speed_type;
toggle_mode: in event port;
throttle_cmd: out data port throttle_data;
end PBA_speed_control;


process implementation PBA_speed_control.basic
subcomponents
    read_data: thread read_data.speed;
    control_laws: thread control_laws.speed;
connections
 C1: port raw_speed -> read_data.raw_data;
 C2: port toggle_mode -> control_laws.toggle;
 C3: port control_laws.cmd -> throttle_cmd;
 C4: port read_data.proc_data -> control_laws.proc_data;
flows none;
modes none;
properties
Period => 20ms;
end PBA_speed_control.basic;

thread read_data
  features
    raw_data: in data port speed_type;
    proc_data: out data port;
end read_data;

thread implementation read_data.speed
end read_data.speed;

thread control_laws
  features
    proc_data: in data port;
    toggle: in event port;
    cmd: out data port throttle_data;
end control_laws;

thread implementation control_laws.speed
calls
  main: {
log_state: subprogram state.log_state;
      };
```

```
end control_laws.speed;

data state
features
log_state: provides subprogram access log_state;
end state;

subprogram log_state
end log_state;
```

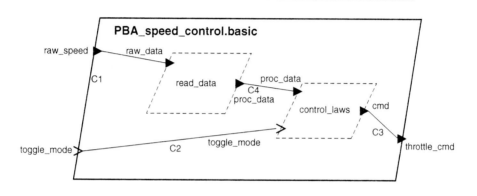

5.6 Summary

This section introduced component types and component implementations to represent the interface specifications of the components that comprise a system and the blueprint of their realizations. Within component type declarations, you establish the externally visible characteristics of components. These characteristics are the name, runtime category, interfaces, properties, variations (modes), and logical flows of the type. You define the internal characteristics of components using component implementation declarations. In declaring a component implementation, you identify the component type that establishes the runtime nature and external characteristics of and a unique name (a combination of the type name and an identifier) for the implementation. Additionally, you define the subcomponents, connections, calls, and variations (if modes are not defined in the type) that comprise the implementation; detail the flows associated with the component type that traverse the various subcomponents; and add or modify properties for the implementation. These implementations establish compositional hierarchies for a system architecture. In doing so, it is possible to establish a complete system hierarchy using exhaustively detailed or partially defined components. As your design progresses and additional design decisions are made, partial definitions can be extended and refined until you have defined a comprehensive declarative system model. You can use this declarative model to generate a complete system instance model.

The next three chapters describe the various categories of AADL components that you use in modeling systems with the AADL.

Chapter 6

Software Components

The AADL application software categories are summarized in Table 6-1. These categories are used to represent the task architecture of the application in terms of processes, thread groups, and threads. They also represent the application software artifacts such as data types, static and local data in terms of a data component, and executable code such as functions and procedures, and source code libraries, in terms of subprograms and subprogram groups.

Table 6-1: *Application Software Components*

Category	Description
thread	Application task as a schedulable unit of concurrent execution
thread group	An abstraction for logically organizing thread, thread group, and data components within a process
process	Protected address space that is enforced at runtime
data	Static or local data in source code and application data types
subprogram	Callable sequentially executable code that represents concepts such as call-return and calls-on methods
subprogram group	An abstraction for organizing subprograms

In addition to the runtime semantics of each category, specific runtime and non-runtime characteristics of application software components are captured in values assigned to properties of those components. For example, information relating to the source code, code descriptions, executable code, and intermediate representations (e.g., object files) are identified by values assigned to the *Source_Text* property for the component and the execution time associated with a thread is defined by the value of the *Compute_Execution_Time* property.

Each software component is summarized in a separate section, where sample declarations are provided. Many of these are minimal. For example, some type declarations do not include any sections and some subcomponent declarations do not include classifier references. This simplicity is intended to provide a focus on the component and, while minimal and not sufficient to describe a realistic architecture, these declarations can be included within a specification as presented. This demonstrates the flexibility of the language in allowing syntactically correct partial specifications.

6.1 Thread

A thread represents an execution path through code that can execute concurrently with other threads. It is a schedulable unit that is assigned (bound) to a processor for execution. The processor schedules the thread and executes the code associated with the thread. The code of a thread executes within the address space defined by the process (see Section 6.3) that contains the thread. You can bind a thread to a virtual processor, which may represent a partition or a particular scheduler, and the thread will execute on the physical processor to which the virtual processor is bound. An AADL thread may be implemented through an operating system thread, or multiple AADL threads may be mapped as logical threads into a single operating system thread (e.g., threads that execute at the same rate). A thread may represent an active object.

A thread can be dispatched repeatedly at fixed time intervals as specified by the *Period* property, referred to as period-based dispatches, or it can be dispatched due to the arrival of data or events on one of the thread's event data or event ports, referred to as port-triggered dispatches. In the latter case, a thread is dispatched immediately upon arrival of a message or event when idle, or it is redispatched once the current execution completes.

The value of the *Dispatch_Protocol* property defines the dispatch characteristics of a thread. The standard dispatch protocol values that have defined semantics are *periodic, aperiodic, sporadic, timed, hybrid,* and *background.*

- *Periodic* threads have a period-based dispatch (a repeated, fixed time interval dispatch) with the assumption that the previous execution has completed before the next dispatch.

- *Aperiodic* threads have a port-based dispatch, which may be queued if the thread is still executing the previous dispatch.

- *Sporadic* threads have a port-triggered dispatch with the additional constraint that the dispatch is delayed if the previous dispatch has occurred in a time less than the time specified by the thread period.

- *Timed* threads are dispatched after a given time specified by the *Period* property unless they are dispatched by arrival of an event or even data, effectively providing a time-out if no input arrives for a given time interval since the last dispatch.

- *Hybrid* threads are dispatched by an event or event data arrival as well as periodically. For periodic, timed, and hybrid threads, a value of the *Period* property for the thread must be declared.

- *Background* threads are dispatched once and execute until completion. In addition to the standard protocols, you can introduce project specific dispatch protocols by modifying the enumeration type *Supported_Dispatch_Protocols* that is found in the *AADL_Project* property set.

Regardless of their dispatch protocol, threads have well-defined runtime states that are built into their execution semantics. The actions for threads are

- *Initialize* for initializing the thread
- *Compute* for nominal execution
- *Recover* for recovering from a fault
- *Activate* to restore thread state on a mode switch where the thread becomes active (executable as part of a given mode)
- *Deactivate* to save thread state on a mode switch where the thread is inactive (not executable as part of a mode)
- *Finalize* for cleanup actions

For each of the thread actions you can specify the application code to be executed using entry point properties. You can specify a reference to source text, name a subprogram classifier, or reference a call sequence to be executed. Details of these properties can be found in Table A-9.

Threads also have a set of runtime states in which they are not performing actions. They are

- *Suspended-D* when a thread is active in the current mode and ready to be dispatched
- *Suspended-M* when a thread is inactive in the current mode and not dispatchable
- *Suspended-P* when the thread execution is preempted by a higher priority thread
- *Suspended-B* when the thread is blocked waiting for a resource locked by another thread
- *Halted* when the thread is stopped and requires initialization before being able to execute

6.1.1 Representation

Table 6-2 shows the graphical and associated textual representations for a thread type and implementation. Two ports have been included for sensor input *sensor_data* and command output *cmd* for the thread.

Table 6-2: *A Sample Thread Implementation with One Subcomponent*

Graphical Representation	AADL Textual Specification

sensor_data → control_laws → cmd	`thread control_laws`
	`features`
	`sensor_data: in data port;`
	`cmd: out data port;`
	`end control_laws;`

static_data	`data static_data` `end static_data;`

control_laws.speed
sensor_data → stability_data → cmd
	`thread implementation`
	`control_laws.speed`
	`subcomponents`
	`stability_data: data`
	`static_data;`
	`end control_laws.speed;`

You can declare a data subcomponent within the thread implementation to represent the internal state of a thread. This state persists between dispatches of the thread and is only accessed by the thread.

You can also indicate that the thread accesses a shared data component from outside the thread by declaring a requires data access feature. In this case, the data component is shared and may require concurrency control (see Section 10.2).

You can express the fact that a thread services remote procedure calls by declaring a provides subprogram access feature for the thread.

6.1.2 Properties

The AADL standard provides a variety of thread-related properties in predeclared property sets (see Section 13.2). They include thread properties such as *Dispatch_Protocol* for indicating the type of thread dispatch, *Dispatch_Trigger* to identify a subset of ports that can cause a dispatch, *Priority* and *Criticality* to indicate the importance of executing a thread over other threads, and properties related to handling of threads in a mode transition. Timing-related properties for threads include thread *Period*, thread *Deadline* and deadlines for thread initialization and finalization, normal execution, error recovery, and activation and deactivation during a mode transition, as well as best-case and worst-case execution time as a time range value. Deployment related properties provide processor binding to map threads to the processor or partition within a processor to execute on. Finally, we have properties related to memory usage such as *Source_Code_Size* and *Source_Data_Size*, as well as properties that identify pieces of source code to be executed by the thread during various stages of its lifetime. For example, the *Initialize_Entrypoint* property identifies the subprogram in the source code for a thread that will execute to initialize the thread.

Listing 6-1 includes sample property associations for a thread type controller where entry points and associated execution times are declared for initialization and nominal executions of the thread. For example, the *Dispatch_Protocol* property value specifies that the thread type *controller* is *Periodic*. Its period is 50 milliseconds, as declared with the Period property. Similarly, the execution time for the thread is a range of 5 milliseconds to 10 milliseconds, as declared with the *Compute_Execution_Time* property. The property association for the property *Allowed_Processor_Binding_Class* declares that instances of the type controller can be bound to any instance of the processor type *marine_certified*.

Listing 6-1: *Sample Thread Properties*

```
thread controller
properties
-- nominal execution properties
Compute_Entrypoint => "control_ep";
Compute_Execution_Time => 5 ms .. 10 ms;
Compute_Deadline => 20 ms;
Dispatch_Protocol => Periodic;
Period => 50 ms;
-- initialization execution properties
Initialize_Entrypoint => "init_control";
Initialize_Execution_Time => 2 ms .. 5 ms;
Initialize_Deadline => 10 ms;
-- binding properties
Allowed_Processor_Binding_Class => (classifier(marine_certified));
end controller;
```

6.1.3 Constraints

Table 6-3 summarizes the component type and implementation declarations that may be contained within a thread type or implementation declaration and identifies the components that may contain a thread. A thread must reside within a process, either as a subcomponent of a process or as a subcomponent of a thread group contained in a process. The thread or the process containing the thread must be bound to

Table 6-3: *Summary of Constraints on Thread Classifier Declarations*

Permitted Content of Thread Classifier Declarations		Thread Permitted as Subcomponent of
Type	**Implementation**	
Features:	Subcomponents:	• Thread group
• Port	• Data	• Process
• Feature group	• Subprogram	• Abstract
• Requires data access	• Subprogram group	
• Provides subprogram access	• Abstract	
• Requires subprogram access	Calls: Yes	
• Provides subprogram group access	Connections: Yes	
• Requires subprogram group access	Flows: Yes	
• Feature	Modes: Yes	
Flows: Yes	Properties: Yes	
Modes: Yes		
Properties: Yes		

memory that is accessible by the processor that executes the code of the thread (i.e., the processor to which the thread is bound).

A thread communicates with other threads or devices through ports, through access to data components shared across multiple threads, and through calls to subprograms serviced by other threads or devices. In addition, a thread may make local calls to subprograms and have persistent local state expressed by data subcomponents. An instance of the subprogram being called may be explicitly modeled as a subcomponent and accessed via a subprogram access connection through the requires subprogram access feature.

6.2 Thread Group

Threads within a process can be organized into a hierarchy of thread groups. This may be desirable if you have a number of threads within a process. Threads within a thread group interact with other threads through the interface of their enclosing thread group. Thread groups can contain data subcomponents that may be shared between threads within the thread group and outside the thread group. Similarly, a thread group may contain subprogram and subprogram group subcomponents that may be called by threads as local calls.

6.2.1 Representations

Table 6-4 is a sample textual specification for a thread group type and implementation declaration. This thread group represents the software controlling the speed of a powerboat. At this time, only the subcomponent declarations of the *control.speed* component implementation are shown (e.g., connections are not included). In this representation, the control algorithms are placed into a thread group separate from a *data_scaling* thread that supports scaling of data and other data management.

The data subcomponent *control_data* represents a persistent data store for control data (e.g., control settings). As a distinct subcomponent, it is explicitly identified as a common store for the *control* thread group—that is, the control data is intended to be accessed by threads within the *control_algorithms* thread group and by the *data_scaling* thread. You must add requires data access features and connections to the thread group and the thread to record this data sharing.

Table 6-4: *A Sample Thread Group Specification*

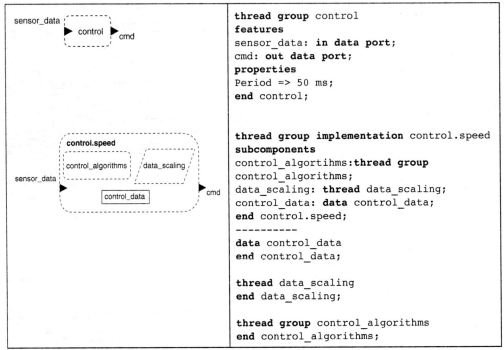

```
thread group control
features
sensor_data: in data port;
cmd: out data port;
properties
Period => 50 ms;
end control;


thread group implementation control.speed
subcomponents
control_algorithms:thread group
control_algorithms;
data_scaling: thread data_scaling;
control_data: data control_data;
end control.speed;
----------
data control_data
end control_data;

thread data_scaling
end data_scaling;

thread group control_algorithms
end control_algorithms;
```

The thread group type declaration *control* includes a property association that defines a *Period* of 50ms. This period applies to all of the periodic threads contained in the thread group, unless specifically overridden by a property association for a particular thread.

6.2.2 Properties

Properties that you declare in thread groups are typically properties that are inherited by the threads contained in a thread group, such as the period or processor binding constraints, if they do not have the appropriate property value. In other words, these property values act as default values for the subcomponents if the property is defined as **inherit**.

6.2.3 Constraints

Table 6-5 summarizes the legal elements within a thread group's type and implementation declarations as well as the components that may

Table 6-5: *Summary of Constraints on Thread Group Classifier Declarations*

Permitted Content of Thread Group Classifier Declarations		Thread Group Permitted as Subcomponent of
Type	Implementation	
Features: • Port • Feature group • Provides data access • Requires data access • Provides subprogram access • Requires subprogram access • Provides subprogram group access • Requires subprogram group access • Feature Flow specifications: Yes Modes: Yes Properties: Yes	Subcomponents: • Data • Subprogram • Subprogram group • Thread • Thread group • Abstract Subprogram calls: No Connections: Yes Flows: Yes Modes: Yes Properties: Yes	• Thread group • Process • Abstract

contain a thread group. A thread group must be contained in a process, either directly or indirectly by being contained in another thread group. A thread group can contain persistent state in the form of static data, expressed by a data subcomponent. This data subcomponent may be accessible to threads and thread groups contained in the given thread group and it can be made accessible outside the thread group.

The features of the thread group type place a constraint on the interactions that subcomponents of the thread group can have. For example, a thread contained in a thread group can only interact with threads outside the thread group through the ports and data access features of the thread group. Similarly, threads in a thread group may make local calls to subprogram and subprogram group subcomponents in the enclosing thread group, process, or system or remote calls to provided subprogram features of another thread.

6.3 Process

An AADL process represents a protected address space that prevents other components from accessing anything inside the process and prevents threads inside a process from causing damage to other processes. In other words, it provides fault isolation for incorrect memory access faults. A process can represent the space partitioning aspect of partitioned architectures such as ARINC653. A process specification does not include an implicit thread; therefore, a complete process specification should contain at least one explicitly declared thread or thread group. This process may have source code associated that represents the executable code and data.

6.3.1 Representations

Table 6-6 contains a partial textual specification for a process and its corresponding graphical representations. The process is shown with examples of three of its allowed subcomponent categories:

1. Thread
2. Thread group
3. Data

Two ports are shown, one as *input* and one as *output* for the process. Since processes represent protected address spaces and often are implemented as separate load images in operating systems, you will usually not provide or require access to data components through the process type—although it is syntactically allowed. Only the subcomponent declarations of the process implementation of *controller.speed* are shown explicitly. Other details of the specification are not included. These

Table 6-6: *Example Representations of a Process*

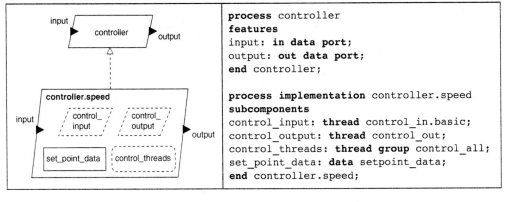

```
process controller
features
input: in data port;
output: out data port;
end controller;

process implementation controller.speed
subcomponents
control_input: thread control_in.basic;
control_output: thread control_out;
control_threads: thread group control_all;
set_point_data: data setpoint_data;
end controller.speed;
```

omissions are legal for a syntactically correct partial specification. This might be useful early in the development where you want to specify the parts your system is composed of without indicating yet how the parts interact with each other. To complete the specification, you add connections that define the interactions between subcomponents and with external components through the process ports.

The left portion of Table 6-6 shows a graphical representation for the textual specification on the right. The subcomponent structure of the process implementation *controller.speed* is shown within its implementation diagram, where individual subcomponents are represented as distinct icons (e.g., the thread icons for the subcomponents *control_input* and *control_output*).

6.3.2 Properties

Standard properties for processes include the specification of the runtime enforcement of memory protection, source code and binary file information, file loading times, and binding constraints. In addition, there are properties that are shared (inherited) by a process's subcomponent threads (e.g., *Period*, *Deadline*, or *Actual_Processor_Binding*). If values are assigned to these properties via a property association within the process type or implementation, that value will be inherited by the subcomponents that share these properties unless overridden by those subcomponents or their classifiers.

6.3.3 Constraints

Table 6-7 summarizes the legal elements within process type and implementation declarations as well as the permitted components that may contain a process. Processes cannot be recursively contained in other processes, but must be contained in system components. Processes can have persistent state in the form of a data subcomponent. This data component can be shared between threads within the process by data access connections to those threads. Data subcomponents can be made accessible to other processes via provides data access features if the underlying operating system supports shared access to memory from multiple process address spaces.

Table 6-7: *Summary of Constraints on Process Classifier Declarations*

Permitted Content of Process Classifier Declarations		Process Permitted as Subcomponent of
Type	Implementation	
Features: • Port • Feature group • Provides data access • Requires data access • Provides subprogram access • Requires subprogram access • Provides subprogram group access • Requires subprogram group access • Feature Flow specifications: Yes Modes: Yes Properties: Yes	Subcomponents: • Data • Subprogram • Subprogram group • Thread • Thread group • Abstract Subprogram calls: No Connections: Yes Flows: Yes Modes: Yes Properties: Yes	• System • Abstract

6.4 Data

You can declare data component types, data component implementations, and data subcomponents to represent data component instances. Data component types represent application data types. Data component implementations represent variants of an application data type and allow you to specify the internal structure of a data type (e.g., the fields of a record). Data component instances can take one of three forms. They can be declared as data subcomponents to represent data (e.g., a global data array with shared access or local data in a subprogram). Alternatively, they can be included in data or event data port declarations to specify the kind of data communicated through the port and accessible to the application source code in the form of a port variable or queue. Finally, they can be included as parameter declarations of subprograms. In all three cases, the type of the data is identified by referring to a data component type or to an implementation of that type.

6.4.1 Representations

Table 6-8 shows sample data type and implementation declarations with their corresponding graphical representations. There are three data

Table 6-8: *Sample Data Component Declarations*

Graphical Representation	AADL Textual Specification
error_data	`data error_data` `end error_data;`

comm_error	`data comm_error`
	`properties`
	`Source_Data_Size => 16 bits;`
	`end comm_error;`

processor_error	`data processor_error`
	`properties`
	`Source_Data_Size => 16 bits;`
	`end processor_error;`

error_data.control
 	
Comm_data
 	
CPU_data
	`data implementation error_data.control`
	`subcomponents`
	`Comm_data: data comm_error;`
	`CPU_data: data processor_error;`
	`end error_data.control;`

types and a data implementation. The *error_data.control* component implementation has two data subcomponents to model elements of a data record. One is an instance of the communication error data type *comm_error*. The other is an instance of the processor error data type *processor_error*. As shown in this example, when declaring data subcomponents it is sufficient to include only a reference to a data type declaration (e.g., in the subcomponent declarations of the data implementation *error_data.control*).

AADL support the declaration of provides subprogram features in data component types. Such a feature can be used to model a method associated with a data type or class in an object-oriented source language.

AADL supports the declaration of component arrays. When applied to data subcomponents we can model data arrays. In other words, AADL can be used as a data modeling language. However, this is not its primary role. You may already use an established data modeling language such as ASN.1 or UML. In this case, you want to map information from the data model that is relevant to the architecture into AADL data components. For that purpose the SAE AADL standards committee has developed a Data Modeling Annex Standard document [DAnnex] that defines a relevant set of properties and common basic data types.

We can declare data components with shared access by multiple threads. This data component can be shared within a process or across processes. We use a *data access* feature declarations to model access to such shared data components (see Section 10.2).

6.4.2 Properties

Among the standard properties for data components are those that enable the specification of source code for the data component, name of the relevant static data variable in the source code, data size, and any concurrency *access* protocol for shared data. For example, in Table 6-8, a size of 16 bits is assigned to the data components *comm_error* and *processor_error* using the *Source_Data_Size* property. This specifies the maximum size of the data represented by the data component. For shared data instances the property *Concurrency_Control_Protocol* defines the protocol for managing access to the shared data.

6.4.3 Constraints

Table 6-9 summarizes what data component type and data component implementation declarations can contain. It also summarizes the permitted components that may contain a data component. Notice that data components can be subcomponents, allowing the specification of

Table 6-9: *Summary of Constraints on Data Classifier Declarations*

Permitted Content of Data Classifier Declarations		Data Permitted as Subcomponent of
Type	**Implementation**	
Features: • Feature group • Requires subprogram access • Requires subprogram group access • Provides subprogram access • Feature Flow specifications: No Modes: No Properties: Yes	Subcomponents: • Data • Subprogram • Abstract Calls: No Connections: Yes Flows: No Modes: No Properties: Yes	• Data • Thread • Thread group • Process • Subprogram • Subprogram group • System • Device • Abstract

Table 6-10: *Summary of Data Classifier References*

Permitted References to Data Classifiers	
In Data Subcomponent	**In Features**
As elements of a data structure in: • Data As sharable data components in: • Threads • Thread group • Process • System • Abstract As local data components in: • Subprograms	To represent data type in: • Data port • Event data port • Requires data access • Provides data access • Parameter

data subfields. As subcomponents of a data implementation, provides subprogram accesses can be used to model methods on objects. A data component can be a subcomponent of a data component representing an element of a data record, of a thread, thread group, process, system, or abstract component representing persistent state in the form of static data, or of a subprogram representing local data.

The constraints on references to data classifiers are shown in Table 6-10. Data classifiers can be referenced by features and by data subcomponent declarations. The reference can be to a data type or a data implementation. Data subcomponent declarations represent static data components whose access can be shared when placed in threads, thread groups, processes, or systems. They represent local data when placed in a subprogram.

6.5 Subprogram

A *subprogram* represents a callable unit of sequentially executable code. Subprogram types define the signature of a subprogram. Subprograms can have parameters through which data values are passed with and returned from a subprogram call. Subprograms can have requires data access features to represent access to persistent state or model passing of parameters by reference. You use outgoing ports to model reporting of exceptions and errors.

Subprogram implementations represent the internals of a subprogram. You are not required to specify a subprogram implementation if there is no need to model the internals of a subprogram. Within a subprogram, data subcomponents represent local variables. We can also specify subprogram call sequences to other subprograms. These calls may be local (the called subprogram is executed in the same thread as that containing the call) or remote (the called subprogram is executed in a different thread than that containing the call).

Subprogram calls can be modeled in two ways. First, you can identify the subprogram to be called by referring to its subprogram classifier. In this case, the linker/loader will provide an instance of the subprogram as part of the execution image of the process. If the call is to be a remote procedure call to a subprogram in another thread, then the *Actual_Subprogram_Call* property is used to identify the thread and the appropriate subprogram access feature of that thread.

Second, a subprogram call can refer to an instance of a subprogram by naming subprogram subcomponents or subprogram access features. In this case, instances of subprograms are declared explicitly as subcomponents and made accessible via subprogram access features and access connections. These subprogram instances may be shared between threads and other components. Threads and other components indicate that they require access to a subprogram or provide access to a subprogram through provides and requires subprogram access features. This supports a component-based development approach, in which it is desirable to completely specify all dependencies to other components, including subprogram call dependencies.

Third, a call can identify a provides subprogram access feature of a data component type to represent method calls for operation on data. The data component being operated on by the method must be passed to the subprogram by value or by reference.

Fourth, a call can identify a requires subprogram access that is connected to a subprogram access feature of another thread to represent remote service/procedure calls.

Finally, a call can identify a provided subprogram access feature of the processor to which an application is bound by referring to the processor's subprogram access feature, in the form **processor** .<subprogram_access_feature name>. This allows you to model service calls to operating system functions provided by the processor.

6.5.1 Representations

Table 6-11 includes example textual and graphical representations relating to the declaration of a subprogram that is a service (method) call for operation on data. It shows the subprogram type *scale_data* and the implementation declaration *scale_data.sensor*. The subcomponent *scale_acc_data* is an instance of *scale_data.sensor* within the *data* component *accelerometer*. The feature *scale_data_access* establishes the *data* component *accelerometer* provides access to an instance of the subprogram implementation *scale_data.sensor*. In this case, access is provided to the subcomponent *scale_acc_data*.

6.5.2 Properties

Standard subprogram properties include declarations relating to the source code for the subprogram such as the *Source_Name*, *Source_Text*, and *Source_Language*. Other properties address the memory size and binding such as *Source_Code_Size*, *Source_Data_Size*, *Source_Stack_Size*, *Source_Heap_Size*, *Allowed_Memory_Binding_Class*, and *Allowed_Memory_ Binding*. In addition, there are properties relating to the calling and execution of the subprogram such as *Actual_Subprogram_Call*, *Allowed_ Subprogram_Call*, *Actual_Subprogram_Call_Binding*, *Allowed_Subprogram_ Call_Binding*, *Subprogram_Call_Type*, *Compute_Execution_Time*, and *Compute_Deadline*.

Table 6-11: *Subprogram Representations*

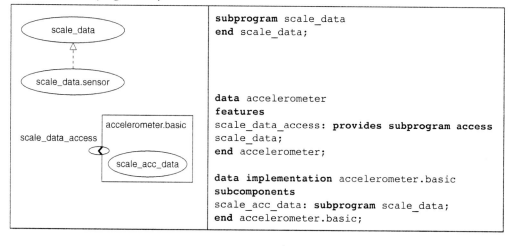

Table 6-12: *Summary of Constraints on Subprogram Classifier Declarations*

Permitted Content of Subprogram Classifier Declarations		Subprogram Permitted as Subcomponent of
Type	**Implementation**	
Features: • Out event port • Out event data port • Feature group • Requires data access • Requires subprogram access • Requires subprogram group access • Parameter • Feature Flow specifications: Yes Modes: Yes Properties: Yes	Subcomponents: • Data • Subprogram • Abstract Subprogram calls: Yes Connections: Yes Flows: Yes Modes: Yes Properties: Yes	• Thread • Thread group • Process • Subprogram group • Subprogram • System • Abstract

6.5.3 Constraints

Table 6-12 summarizes the permitted elements of a subprogram's component type and implementation declarations. Subprograms can contain data subcomponents to represent local data. Subprograms cannot contain instances of other subprograms; these must be declared in threads, thread groups, processes, or systems and be made available through subprogram access features. Note that AADL does not require you to explicitly model subprogram instances. Instead, you can refer to subprogram classifiers in subprogram calls to identify the subprogram signature.

6.6 Subprogram Group

You can use subprogram groups to represent a collection of callable routines such as a subprogram library by declaring a subprogram group type with provides subprogram access features. Similar to subprograms, you can model software libraries by subprogram group types, leaving the job of determining the number of copies of the library to the linker/loader; or you can explicitly declare instances of such libraries through subprogram group subcomponent declarations. In subprogram calls you refer to the subprogram group type or to the subprogram group instance and then the provides subprogram access feature (e.g., *filter_library.basic_acc*).

6.6.1 Representations

Examples of subprogram group declarations and corresponding graphical representations are shown in Table 6-13. A subprogram group is represented graphically as an oval with a dashed border. The subprogram group type *filter_library* includes the provides subprogram access features: *basic_acc, lowP_filter,* and *hiP_filter*. The subprogram group implementation *filter_library.BLH* contains the subprogram instances: *basic_filter, LowP_filter,* and *HiP_filter* that correspond to the access

Table 6-13: *Example Subprogram Group Declarations*

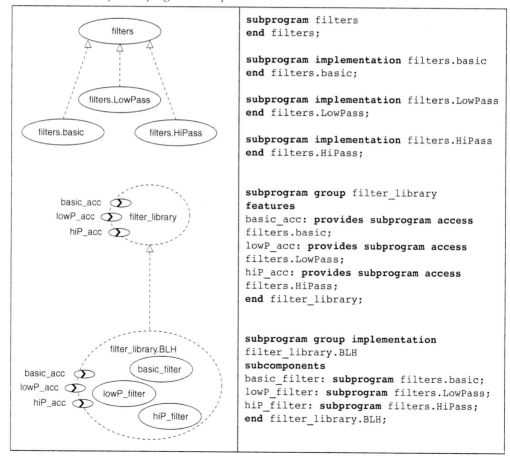

```
subprogram filters
end filters;

subprogram implementation filters.basic
end filters.basic;

subprogram implementation filters.LowPass
end filters.LowPass;

subprogram implementation filters.HiPass
end filters.HiPass;


subprogram group filter_library
features
basic_acc: provides subprogram access
filters.basic;
lowP_acc: provides subprogram access
filters.LowPass;
hiP_acc: provides subprogram access
filters.HiPass;
end filter_library;


subprogram group implementation
filter_library.BLH
subcomponents
basic_filter: subprogram filters.basic;
lowP_filter: subprogram filters.LowPass;
hiP_filter: subprogram filters.HiPass;
end filter_library.BLH;
```

features of the type *filter_library*. For example, the feature *basic_acc* provides access to an instance of the subprogram *filter_library.BLH*, which is the subcomponent *basic_filter*.

6.6.2 Properties

The standard properties of subprogram groups are associated with the properties of subprograms. All subprogram properties that are declared as **inherit** can be declared within a subprogram group declaration. Those values declared in the subprogram group apply to all of the subprograms within the group unless that value is overridden by a property association elsewhere (e.g., within an individual subprogram instance).

6.6.3 Constraints

Table 6-14 summarizes the permitted elements of a subprogram group's type and implementation declarations. You may indicate that a subprogram library uses another library by specifying a requires subprogram group access feature. You may choose to model the subprograms contained in a subprogram library by declaring them as subprogram subcomponents in the subprogram group implementation.

Table 6-14: *Summary of Constraints on Subprogram Group Classifier Declarations*

Permitted Content of Subprogram Group Classifier Declarations		Subprogram Group Permitted as Subcomponent of
Type	Implementation	
Features: • Feature group • Provides subprogram access • Requires subprogram access • Requires subprogram group access • Provides subprogram group access • Parameter Flow specifications: No Modes: No Properties: Yes	Subcomponents: • Subprogram • Subprogram group • Data • Abstract Subprogram calls: No Connections: subprogram access Flows: No Modes: No Properties: Yes	• Subprogram group • Thread • Thread group • Process • System • Abstract

Chapter 7

Execution Platform Components

Execution platform components represent the resources of the computer system as well as elements of the external physical environment. These component categories are summarized in Table 7-1.

The physical resources of the computer system are represented by the processor, memory, and bus categories. For example, a processor component represents a CPU chip or a processor board including operating system functionality. Similarly, a bus component represents an

Table 7-1: *Execution Platform Categories*

Category	Description
processor	Schedules and executes threads and virtual processors
virtual processor	Used to model hierarchical schedulers, partitions, and virtual machines
memory	Stores data and code
bus	Interconnects processors, memory, and devices
virtual bus	Used to model communication abstractions such as a virtual channel or communication protocols
device	Represents sensors, actuators, or other entities of the external physical environment

onboard bus such as PCI, or a network such as an Ethernet or CAN bus. It may include the functionality of a protocol. Bus components are used to physically interconnect execution platform components. A virtual processor can represent a scheduler or a virtual machine such as a Java virtual machine. A virtual bus can represent a virtual channel or a communication protocol. Memory components can represent working memory such as RAM or cache memory as well as persistent storage such as a hard drive. A device component can represent entities that interface with the external environment, such as a speed sensor or a switch actuator, or they can represent an entity of the external environment such as an engine with sensors and actuators represented as ports. Finally, we can combine these execution platform components into system components to represent computer platforms such as a board or cabinet, to physical subsystems or systems such as an engine.

7.1 Processor

A processor represents the hardware and associated software that is responsible for scheduling and executing threads. The scheduler that is part of the processor abstraction may be an interrupt handler or operating system supporting specific scheduling protocols. The processor abstraction may also include higher-level operating system services that are made accessible to the application through ports and subprogram access features. A processor may include memory for code and data as subcomponents. Alternatively, code and data may be stored in an external memory component that is connected to the processor via a bus component. The mapping of software onto processors is declared using binding properties.

7.1.1 Representations

Table 7-2 shows a type and implementation declaration for a processor including both textual and corresponding graphical representations. In this example, a single processor system with memory contained inside of the processor is shown. This memory is considered as directly accessible by the processor.

In the textual representation, the properties subclauses define the hardware description language (*Hardware_Source_Language*) and the files that contain the source code for the hardware description

Table 7-2: *A Sample Processor Textual and Graphical Representation*

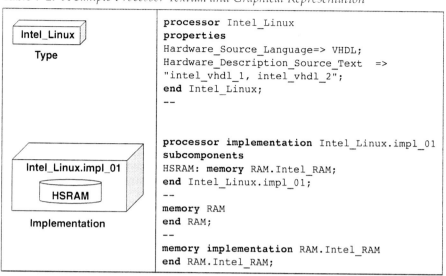

Intel_Linux
Type
	```processor Intel_Linux```
	```properties```
	```Hardware_Source_Language=> VHDL;```
	```Hardware_Description_Source_Text =>```
	```"intel_vhdl_1, intel_vhdl_2";```
	```end Intel_Linux;```
	```--```

Intel_Linux.impl_01
HSRAM
Implementation
	```processor implementation Intel_Linux.impl_01```
	```subcomponents```
	```HSRAM: memory RAM.Intel_RAM;```
	```end Intel_Linux.impl_01;```
	```--```
	```memory RAM```
	```end RAM;```
	```--```
	```memory implementation RAM.Intel_RAM```
	```end RAM.Intel_RAM;```

(*Hardware_Description_Source_Text*). The processor implementation declaration of *Intel_Linux.impl_01* includes a single memory subcomponent *HSRAM*. The memory subcomponent's type and implementation declarations are shown.

The corresponding graphical representations of type and implementation are shown to the left of the textual representation in Table 7-2. The nesting of the memory graphic (labeled *HSRAM*) within the processor graphic shows containment.

If the processor does not contain internal memory, it must have access to external memory components via a bus. In this case, the bus used to access the memory must either be declared as a processor subcomponent or made accessible outside the processor via a provides bus access feature, or the processor must require access to a bus external to the processor.

A processor may support a number of different communication protocols. You can model this by declaring virtual bus subcomponents to satisfy protocol requirements by port connections. Alternatively, you can specify the set of protocols supported by a processor through the *Provided_Virtual_Bus_Class* property by naming the respective virtual bus classifiers as the property value. The *Required_Virtual_Bus_Class* indicates for a port or port connection, which protocol is required.

7.1.2 Properties

The following standard processor properties can be used in a processor declaration. The example from Table 7-2 showed hardware description properties. A *Scheduling_Protocol* property indicates the protocol used to schedule threads. An *Allowed_Dispatch_Protocol* property specifies the dispatch protocols supported by the processor. We have properties that indicate the speed of the processor (*Clock_Period, Assign_Time, Scaling_Factor*) and the cost of context switch time between threads within the same process and between processes, and properties that identify the memory requirements of any software associated with the processor.

7.1.3 Constraints

Table 7-3 summarizes the legal elements within a processor type and implementation declaration as well as the permitted components that may contain a processor.

Table 7-3: *Summary of Constraints on Processor Classifier Declarations*

Permitted Content of Processor Classifier Declarations		Processor Permitted as Subcomponent of
Type	Implementation	
Features: • Provide subprogram access • Provides subprogram group access • Port • Feature group • Requires bus access • Provides bus access • Feature Flow Specifications: Yes Modes: Yes Properties: Yes	Subcomponents: • memory • bus • virtual processor • virtual bus • abstract Calls: No Connections: Bus access Flows: Yes Modes: Yes Properties: Yes	• System • Abstract

7.2 Virtual Processor

A virtual processor represents a logical resource for scheduling and executing software. You can use it to represent a virtual machine such as the Java VM, a partition of a processor, or a scheduler in a scheduler hierarchy. You bind virtual processors to processors or to other virtual processors that eventually are bound to processors. You can also declare virtual processors as a subcomponent of a physical processor or a subcomponent of a virtual processor that is contained in a processor.

7.2.1 Representations

Example virtual processor declarations and graphical representations are shown in Table 7-4. The virtual processor graphical icon is the processor icon with dotted boundaries. There is one implementation of the virtual processor type *custom* shown in Table 7-4. This implementation includes a virtual bus as a subcomponent. Notice the assignment of 5 nanoseconds to the *Startup_Execution_Time* property in the type declaration. This value will apply to all instances of the type unless overridden for that instance.

Virtual processors describe how processors are divided into multiple logical resources, each with a guaranteed portion of the physical processor. For example, the ARINC 653 partition concept is represented by processes that are bound to virtual processors. The processes specify

Table 7-4: *Virtual Processor Representations*

custom	```virtual processor custom properties Startup_Execution_Time => 5 ns; end Custom; -- ```
custom.secure sbus12	```virtual processor implementation custom.secure subcomponents sbus12: virtual bus secure.bit_12; end custom.secure; -- virtual bus secure end secure; -- virtual bus implementation secure.bit_12 end secure.bit_12;```

the need for space partitioning through protected address spaces, while the virtual processors assure time partitioning of the processor as resource. Virtual processors can be used to model different language runtime systems on the same processor with different application programming interfaces and functionality. In addition, virtual processors can represent multiple schedulers including hierarchical schedulers, with different scheduling protocols. For example, a virtual processor can be used to represent the Ada language runtime system with its task execution model. Similarly, a virtual processor can be used to represent an operating system thread that schedules and executes application tasks with the same period—represented by AADL threads—using a cyclic executive scheduling protocol such as a thread calling individual tasks as functions.

7.2.2 Properties

Virtual processors share properties with a processor, except for properties that describe the hardware aspects of a processor (e.g., hardware description, data movement in memory, and hardware clock properties) and those specific to a virtual processor. Virtual processor specific properties include properties that address the mapping of a virtual processor to a processor such as *Allowed_Processor_Binding* as well as the *Startup_Execution_Time* property that specifies the execution time and dispatch characteristics associated with a virtual processor. You can indicate the desired quality of service of a protocol using the *Required_Connection_Quality_Of_Service* property to indicate protocol service needs such as guaranteed delivery instead of prescribing a specific protocol. Similarly, the *Required_Connection_Quality_Of_Service* property associated with a virtual bus indicates the provided service level.

7.2.3 Constraints

Table 7-5 summarizes the legal elements within virtual processor type and implementation declarations as well as the permitted components that may contain a virtual processor.

Table 7-5: *Summary of Constraints for Virtual Processor Declarations*

Permitted Content of Virtual Processor Declarations		Virtual Processor
Type	Implementation	Permitted as Subcomponent of
Features: • Provides subprogram access • Provides subprogram group access • Port • Feature group • Feature Flow Specifications: Yes Modes: Yes Properties: Yes	Subcomponents: • Virtual processor • Virtual bus • Abstract Calls: No Connections: Yes Flows: Yes Modes: Yes Properties: Yes	• Processor • Virtual processor • System • Abstract

7.3 Memory

Memory represents storage components for data and executable code. Memory components include randomly accessible physical storage (e.g., RAM, ROM), reflective memory, or permanent storage such as disks. Binding properties define the memory components in which software components are stored.

7.3.1 Representations

An example memory declaration and its graphical representation are shown in Table 7-6. In this example, a memory of the type *RAM* is declared with a single feature *bus01* that establishes that all instances of *RAM* require access to the bus *membus*. No explicit properties for this type are declared. The type and implementation declarations for the requires bus access to *bus01* are shown at the end of the listing.

The memory implementation *RAM.compRAM* declares that this implementation of the memory type *RAM* includes memory subcomponents *HSRAM* and *SRAM*. The subcomponents of the memory implementation *RAM.compRAM* are declared as implementations of a common type *XRAM*. An expanded memory composition can be used to model a complicated memory bank. These examples show how memory can contain other memory components and that memory must be connected to a bus unless it is enclosed in a processor.

Table 7-6: *A Sample Memory Textual and Graphical Representation*

| | ```
memory RAM
features
bus01: requires bus access memory_bus;
end RAM;
--
bus memory_bus
end memory_bus;


--

memory implementation RAM.compRAM
subcomponents
HSRAM: memory XRAM.HSRAM;
SRAM: memory XRAM.SRAM;
end RAM.compRAM;
--
memory XRAM
end XRAM;
--
memory implementation XRAM.HSRAM
end XRAM.HSRAM;
--
memory implementation XRAM.SRAM
end XRAM.SRAM;
``` |

Typically, you map two kinds of software components to memory components: processes and data components. For example, a process has memory requirements for code, local data in form of stack, and dynamic data in form of heap, expressed by the *Source_Code_Size*, *Source_Stack_Size*, and *Source_Heap_Size* properties. A data component represents static data that persists between thread dispatches and may be accessible by more than one thread. The *Source_Data_Size* property indicates the memory requirements for an instance of a given data component type (i.e., for such static data components) as well as for data communicated through ports.

Memory binding properties, as discussed in the software component sections, specify the mapping of the application components to memory components.

7.3.2 Properties

Since they have a physical runtime presence, memory components have properties such as word size and word count. Standard memory properties include
```

- memory access protocol
- read and write times for memory access
- memory size in terms of word size and count
- base address to indicate the physical starting address of the memory unit

The default value for memory access (*Memory_Protocol*) is read–write but memory access can be assigned as read only or write only.

7.3.3 Constraints

Table 7-7 summarizes the legal elements within memory type and implementation declarations as well as the permitted components that may contain a memory component. An individual memory component must be contained in a processor, declared as a subcomponent of a memory unit, or connected to a processor through a bus.

A memory component can represent memory inside of a processor or as a separate execution platform unit that is connected to the processor via a bus. Memory banks can be modeled abstractly as a single memory component, or explicitly as a composite memory unit by placing memory subcomponents inside a memory component implementation.

Table 7-7: *Summary of Constraints on Memory Classifier Declarations*

Permitted Content of Memory Classifier Declarations		Memory Permitted as Subcomponent of
Type	Implementation	
Features: • Requires bus access • Provides bus access • Feature group • Feature Flow specifications: No Modes: Yes Properties: Yes	Subcomponents: • Memory • Bus • Abstract Subprogram calls: No Connections: Bus access Flows: No Modes: Yes Properties: Yes	• Processor • Memory • System • Abstract

7.4 Bus

A bus represents the physical hardware and associated communication protocols required to support the interactions among physical components. Examples of buses are PCI bus, VME bus, CAN bus, Ethernet, and wireless networks. The role of a bus in supporting communication is defined using bus access features (that declare the need for a physical connection) and bus access connections (that declare a physical connection).

Buses are typically used to connect computer hardware components. We can also use the bus component to represent physical resources such as electrical power that is distributed to physical components and computer hardware components. We use bus access features (see Section 10.3) to indicate the need of a resource such as electrical power or the ability to contribute electrical power via a set of properties defining for modeling physical resources.

7.4.1 Representations

Table 7-8 shows a portion of an AADL textual specification and corresponding graphical representation. Included in the example are a processor type declaration for *Intel_Linux* and two bus type declarations for *memory_bus* and *network_bus*. The processor type declaration for *Intel_Linux* includes a requires bus access declaration for the bus type *memory_bus* and the bus type declaration *memory_bus* includes a requires bus access for the bus type *network_bus*. These required accesses are shown in the graphic on the left side of Table 7-8.

Table 7-8: *A Sample Bus Specification: Textual and Graphical Representations*

memory_bus
□ NBA
	`bus memory_bus`
	`features`
	`NBA: requires bus access  network_bus;`
	`end memory_bus;`

network_bus	`bus network_bus` `end network_bus;`

Intel_Linux
□ MBA
	`processor Intel_Linux`
	`features`
	`MBA: requires bus access memory_bus;`
	`end Intel_Linux;`

For example, a processor and memory require access to a bus to enable the processor to utilize the memory or to communicate with a camera via a USB bus. The need for bus access is defined in the type declarations for the processor and memory and the connections to the bus are declared explicitly within the system implementation that contains the processor, memory, and bus. Buses can be connected directly to other buses to represent inter-network communication. This is specified by requires bus access features within the bus type and explicit connection declarations between instances of these buses.

7.4.2 Properties

There are a number of standard properties that specify important bus characteristics. These include connection (e.g., *Allowed_Connection_Type* that specifies the categories of connections that are supported), message size (e.g., *Allowed_Message_Size*), transmission time (e.g., *Transmission_ Time*), hardware source language descriptions (e.g., *Hardware_Source_ Language*), and data movement time (e.g., *Transmission_Time* that specifies the time required to move a block of bytes on a bus) properties. You can use the *Provided_Virtual_Bus_Class* property to indicate the protocols supported by the bus type.

7.4.3 Constraints

Table 7-9 summarizes the legal elements within bus type and implementation declarations as well as the permitted components that may

Table 7-9: *Summary of Constraints on Bus Classifier Declarations*

Permitted Content of Bus Classifier Declarations		Bus Permitted as Subcomponent of
Type	**Implementation**	
Features: • Requires bus access • Feature group • Feature Flow specifications: No Modes: Yes Properties: Yes	Subcomponents: • Virtual bus • Abstract Subprogram calls: No Connections: Yes Flows: No Modes: Yes Properties: Yes	• Processor • Memory • Device • System • Abstract

contain a bus component. A bus may contain virtual bus subcomponents to indicate the protocols supported by the bus as an alternative to the *Provided_Virtual_Bus_Class* property. You use the requires bus access to indicate that an instance of this may be connected to another bus (e.g., a local network may be connected to a backbone).

7.5 Virtual Bus

A virtual bus component represents a logical abstraction such as a virtual channel or communication protocol. Virtual buses can support connections between application software components on the same processor or across multiple processors via buses.

7.5.1 Representations

The graphical and textual representations of virtual bus type and implementation declarations are shown in Table 7-10. In the example, a single implementation of the virtual bus type *high_speed_bus* is shown. In addition, a virtual bus can be a subcomponent of a processor or virtual processor as shown in Table 7-4.

A virtual bus can be declared as a subcomponent of a processor or bus or it can be bound to a processor or bus. Virtual buses can require other virtual buses, allowing the modeling of protocol hierarchies. Virtual buses can be used to model virtual "channels" that represent a partition of the bandwidth of a network or a bus as well as specific communication protocols and protocol stacks. The *Provided_Virtual_Bus_Class* property can be used to indicate that a protocol is supported. For example, a processor may indicate that it supports a certain set of protocols by referring to the virtual bus types that represent the supported protocols. Similarly, a bus may indicate the protocols it supports. A connection between ports may indicate that it requires a particular protocol using the *Required_Virtual_Bus_Class* property.

Table 7-10: *Virtual Bus Declarations*

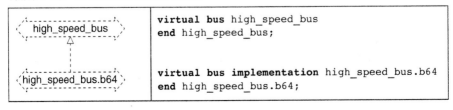

high_speed_bus	`virtual bus high_speed_bus` `end high_speed_bus;`
high_speed_bus.b64	`virtual bus implementation high_speed_bus.b64` `end high_speed_bus.b64;`

A protocol stack can be modeled by a single virtual bus type that indicates all the protocols of the protocol stack it makes available for application port connections through the *Provided_Virtual_Bus_Class* property. The same stack can also be modeled by each virtual bus type that represents a protocol indicating that it requires other protocols. The collection of *Required_Virtual_Bus_Class* properties in the different virtual bus types effectively represents the protocol stack hierarchy.

7.5.2 Properties

As with a virtual processor, the virtual bus abstraction shares a number of properties with its hardware counterpart. However, properties that are not common include hardware specific properties and data movement related properties. Virtual buses can specify that they require a specific protocol through the *Required_Virtual_Bus_Class* property. Desired virtual bus quality of service characteristics such as secure delivery or guaranteed delivery can be specified through the *Required_Connection_Quality_Of_Service* property. You can also declare that a virtual bus provides such characteristics using the *Provided_Connection_Quality_Of_Service* property.

7.5.3 Constraints

The legal elements within bus type and implementation declarations as well as the permitted components that may contain a virtual bus component are summarized in Table 7-11. Virtual buses must be contained in or bound to physical processors or buses.

Table 7-11: *Summary of Constraints on Virtual Bus Classifier Declarations*

Permitted Content of Virtual Bus Classifier Declarations		Virtual Bus Permitted as Subcomponent of
Type	**Implementation**	
Features: • None Flow specifications: No Modes: Yes Properties: Yes	Subcomponents: • Virtual bus • Abstract Subprogram calls: No Connections: No Flows: No Modes: Yes Properties: Yes	• Processor • Virtual processor • Bus • Virtual bus • System • Abstract • Device

7.6 Device

Device abstractions represent entities that interface with or are part of the external environment of an embedded software system, such as sensors and actuators, or engines, Global Positioning System (GPS), or cameras. Devices can represent a physical entity or a software simulation of a physical entity. A device may have its functionality fully embedded in the hardware, or it may include a device driver that must execute on a processor. The device driver and its binding can be characterized by device-specific properties to indicate that the driver is to be executed as part of the operating system. Alternatively, an explicit device driver thread can be declared to model the fact that the device driver executes separately from the OS kernel.

A device can interact with other application components through ports or subprogram access features. Device ports may represent registers or ports of a device driver. Devices interact with hardware components through bus access connections.

7.6.1 Representations

Table 7-12 shows an excerpt from an AADL model that describes a *device speed_sensor* interacting through a bus with a processor *speed_processor*. The processor executes the device driver for the *speed_sensor*. The requirement for bus access is specified in the type declaration for *speed_sensor*. Similarly, the need for bus access is declared within the processor type declaration for *speed_processor*. Notice that the out data port declared on the *speed_sensor* device type provides the rate data from the sensor. A device also can be used to represent a more complex physical element, such as an engine where the ports can represent the engine's sensors and actuators.

A device can have a physical connection to a processor via a bus as well as logical connections through ports to application software components. As with all logical connections among components residing on distinct execution platform elements, these logical connections must be supported by (be bound to) one or more buses that physically connect the device to the processor running the application thread. A device may require driver software that is executed on an external processor. A device's external driver software may be considered part of a processor's execution overhead, or it may be explicitly declared as a thread with its own execution properties.

Table 7-12: *A Sample Device Specification: Textual and Graphical Representation*

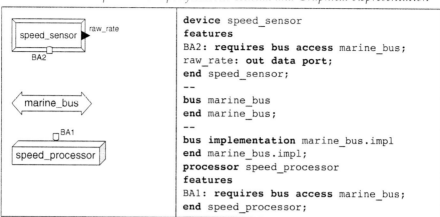

```
device speed_sensor
features
BA2: requires bus access marine_bus;
raw_rate: out data port;
end speed_sensor;
--
bus marine_bus
end marine_bus;
--
bus implementation marine_bus.impl
end marine_bus.impl;
processor speed_processor
features
BA1: requires bus access marine_bus;
end speed_processor;
```

7.6.2 Properties

Standard device properties encompass the dual software and hardware character of a device and include software-specific properties such as the names of source code files, source code language, code size, and properties for execution platform binding. Other properties are related to the execution platform (hardware) aspects of a device. These include properties specifying the files that contain the hardware description language for the device and the language used for that description. In addition, there are properties for specification of the thread properties of any device software that is executing on a processor, such as dispatch protocols and execution time-related properties.

You may want to explicitly model the implementation of device drivers. In that case, you define an abstract component implementation that contains an instance of the device driver subprogram or an instance of the device driver as thread. This classifier is then associated with the device using the *Implemented_As* property.

7.6.3 Constraints

Table 7-13 summarizes the legal elements within device type and implementation declarations as well as the permitted components that may contain a device. A device may support communication protocols as indicated by virtual bus subcomponents or the *Provided_Virtual_ Bus_Class* property. A device may also contain a bus through which other components can be connected to the device, or the device may require access to an externally provided bus.

Table 7-13: *Summary of Constraints on Device Classifier Declarations*

Permitted Content of Device Classifier Declarations		A Device Permitted as Subcomponent of
Type	**Implementation**	
Features: • Port • Feature group • Provides subprogram access • Provides subprogram group access • Requires bus access • Provides bus access • Feature Flow specifications: Yes Modes: Yes Properties: Yes	Subcomponents: • Bus • Virtual bus • Data • Abstract Subprogram calls: No Connections: Yes Flows: Yes Modes: Yes Properties: Yes	• System • Abstract

Chapter 8

Composite and Generic Components

In this section, we discuss the system and abstract component categories that provide a framework for specifying composite components and generic components. Systems are used to define the composite components of a system recursively in terms of system subcomponents as well as specific software, computer hardware, and physical system components. Abstract components can be used to represent a component hierarchy as composites and individual components that are generic. Abstract components can be refined later in the design process to declare explicitly the runtime category of the components that comprise the runtime architecture. Components contained in a system component or abstract component can only interact with external components through the features that make up the interface of a system or abstract component.

8.1 System

The system abstraction represents a composite that can include software, execution platform, or system components. It can be used to represent composites of computer hardware components (e.g., a processor board or cabinet), a collection of application software processes (e.g., a flight navigation system), a combination of software and hardware components (e.g., an air data computer), collections of physical system components (e.g., a cockpit consisting of various panels, displays, flight sticks, and other controls), or a composite of subsystems or a system of systems (e.g., an aircraft system). In addition, models consisting of only system components can be used as a generic representation of a component-based conceptual view of a system architecture.

8.1.1 Representations

A system can consist of various combinations of software, execution platform, and system components. For example, a system may consist only of software (processes, subprograms, subprogram groups, or data—threads or thread groups must be contained in processes) or only of computer system components (processor, bus, memory, and systems of these), components of the external environment (devices, buses, and their composites in form of systems), as well as combinations of such systems.

Listing 8-1 provides textual and graphical representations of an implementation of the *system* type *integrated_control*. Subcomponent declarations are shown in the implementation. However, the requisite classifier declarations are not shown (e.g., the process type declaration for the *process* type *controller*). In the graphical portrayal, the subcomponents of *integrated_control.PBA* are shown.

Listing 8-1: *A Sample System Specification: Textual and Graphical Representation*

```
system integrated_control
end integrated_control;
--
system implementation integrated_control.PBA
subcomponents
control_process: process controller.speed_control;
set_point_data: data set_points;
navigation_system: system core_system.navigation;
real_time_processor: processor rt_fast.rt_processor;
hs_memory: memory rt_memory.high_speed;
high_speed_bus: bus network_bus.HSbus;
end integrated_control.PBA;
```

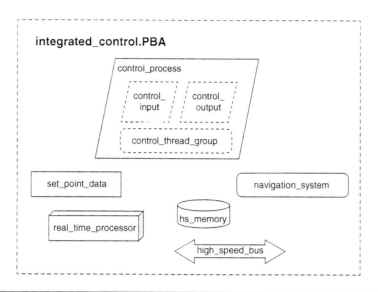

8.1.2 Properties

Many standard properties that can be attached to system components are properties that are shared with contained components as inherited properties. Some properties are not inherited and apply to the system component itself, such as *Startup_Execution_Time* and *Startup_Deadline*. It is common for you to introduce additional properties for characteristics that are relevant at the system level, such as weight, reliability, or security.

8.1.3 Constraints

Table 8-1 summarizes the legal elements within system type and implementation declarations as well as the permitted components that may contain a system. Notice that a system cannot contain a thread or thread group directly; they must be contained in a process.

8.2 Abstract

The abstract component category is a generic category that allows you to declare component types and implementations without choosing their runtime category. With this category, you can develop

- Conceptual component-based views of a system architecture
- Generic component-based reference architectures

Table 8-1: *Summary of Constraints on System Classifier Declarations*

Permitted Content of System Classifier Declarations		A System Permitted as Subcomponent of
Type	**Implementation**	
Features: • Port • Feature group • Provides data access • Provides subprogram access • Provides subprogram group access • Provides bus access • Requires data access • Requires subprogram access • Requires subprogram group access • Requires bus access • Feature Flow specifications: Yes Modes: Yes Properties: Yes	Subcomponents: • Data • Subprogram • Subprogram group • Process • Processor • Virtual processor • Memory • Bus • Virtual bus • Device • System • Abstract Subprogram calls: Yes Connections: Yes Flows: Yes Modes: Yes Properties: Yes	• System • Abstract

- Architecture templates or patterns that can be instantiated and applied to an architecture (e.g., a redundancy pattern that is applied as an aspect)

As needed, you can refine these generic declarations into a thread, thread group, process, system, data, subprogram, processor, virtual processor, memory, bus, virtual bus, or device. This technique is especially useful in the early phases of development by providing alternatives for architectural trade-off studies and representations for analysis (e.g., system-wide latency analyses can be conducted without having to define the runtime nature of the components involved).

8.2.1 Representations

The basic forms for abstract component type and implementation declarations are the same as for the runtime-specific component categories. An example is shown in Listing 8-2 that is similar to the system declaration shown in Listing 8-1 but, in addition, includes an abstract subcomponent *error_response*. The graphical representation associated with the specification is presented in the middle section of the listing. In the lower portion, the declarations are shown that extend the abstract component implementation to the system category. The result is an equivalent description of the system component presented in Listing 8-1.

Listing 8-2: *A Sample Abstract Specification: Textual and Graphical Representation*

```
abstract integrated_control
end integrated_control;
--
abstract implementation integrated_control.PBA
subcomponents
control_process: process controller.speed_control;
set_point_data: data set_points;
navigation_system: system core_system.navigation;
real_time_processor: processor rt_fast.rt_processor;
hs_memory: memory rt_memory.high_speed;
high_speed_bus: bus network_bus.HSbus;
error_response: abstract error_manager.real_time;
end integrated_control.PBA;

system integrated_control_sys
extends integrated_control
end integrated_control_sys;

system implementation integrated_control_sys.PBA
extends integrated_control.PBA
end integrated_control_sys.PBA;
```

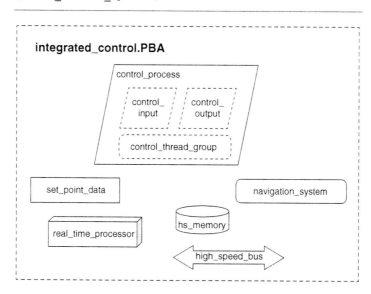

8.2.2 Properties

All of the properties available to the other categories apply to an abstract component. However, when an abstract component is refined to a non-abstract category, only those properties that apply to the non-abstract category are valid.

8.2.3 Constraints

Table 8-2 summarizes the legal elements within abstract type and implementation declarations as well as the permitted components that may contain an abstract component.

Table 8-2: *Summary of Constraints on Abstract Classifier Declarations*

Permitted Content of Abstract Classifier Declarations		Abstract Permitted as Subcomponent of
Type	**Implementation**	
Features: • Port • Feature group • Provides data access • Provides subprogram access • Provides subprogram group access • Provides bus access • Requires data access • Requires subprogram access • Requires subprogram group access • Requires bus access • Feature Flow specifications: Yes Modes: Yes Properties: Yes	Subcomponents: • Data • Subprogram • Subprogram group • Thread • Thread group • Process • Processor • Virtual processor • Memory • Bus • Virtual bus • Device • System • Abstract Subprogram calls: Yes Connections: Yes Flows: Yes Modes: Yes Properties: Yes	• Subprogram group • Data • Subprogram • Thread • Thread group • Process • Processor • Virtual processor • Memory • Bus • Virtual bus • Device • System • Abstract

Chapter 9

Static and Dynamic Architecture

Systems have a static and a dynamic architecture. The static architecture is represented by a collection of components that are structured in a component hierarchy. For each component in the hierarchy, its component implementation defines the internal structure in terms of subcomponents and interactions in terms of connections. The dynamic architecture is represented by a collection of component and connection configurations that are controlled by modes. Modes effectively represent operational modes and fault tolerant configurations of both software and hardware, as well as modal behavior of individual components. In this chapter, we discuss subcomponents and arrays of subcomponents to represent the static architecture, as well as modes and mode transitions to represent the dynamic architecture. Component interactions in terms of connections and information flows throughout the system are discussed in subsequent chapters.

9.1 Subcomponents

A subcomponent declaration defines a component instance by referencing a component type or implementation. This reference identifies a pattern for the subcomponent that defines the category and characteristics of that instance.

9.1.1 Declaring Subcomponents

The template for a subcomponent declaration for a subcomponent follows:

```
name : <component category> [component reference]
       [{properties}] [in modes];
```

Both a subcomponent name and a component category are required. The entry <component category> must be replaced by one of the legal AADL component categories, such as thread, process, memory, or abstract. The name of a subcomponent is a single AADL identifier. This identifier must be unique within the component implementation (i.e., you cannot have two subcomponents with the same name, or a connection and a subcomponent with the same name).

The component reference identifies a component type or component implementation whose instance the subcomponent represents. The reference can be to a component type or implementation within the same package (e.g., *core_system*) or in a different package (e.g., *data_pkg::set_points*). The subcomponent name can be the same as the name of the component type in the component reference. The component reference is optional for a partially complete component implementation. Similarly, property associations are optional. Finally, you only need to declare **in modes** if the subcomponent is part of a component implementation with modes and the subcomponent is only active in certain modes. Example subcomponent declarations are shown in Listing 9-1.

Listing 9-1: *Example Subcomponent Declarations*

```
control_process: process controller.speed
    in modes (init, operational);
set_point_data: data data_pkg::set_points;
navigation_system: system core_system;
real_time_processor: processor rt_fast.rt_processor;
marine_bus: bus hardware_pkg::marine_bus;
marine_certified_RAM: memory { Memory_Protocol  => read_write;};
init_process: process init_process in modes (init)
```

9.1.2 Using Subcomponent Declarations

Subcomponents are declared within the subcomponents section of a component implementation and define component instances that are contained within that implementation. Since a contained component may itself contain subcomponents, subcomponent declarations can be used to establish a compositional component hierarchy. For example, a control system component can contain a system component that itself contains a directional control system. The directional control system can contain a process, which contains a thread that provides the control law algorithm processing. This is shown in Listing 9-2, where the system implementation *basic.speed_control* contains three subcomponents. The subcomponent declaration of *speed_control* references the process implementation *control.speed*, which contains two thread subcomponents.

Listing 9-2: *Defining a Component Hierarchy with Subcomponent Declarations*

```
system implementation basic.speed_control
  subcomponents
    speed_sensor: device sensor.speed;
    throttle: device actuator.speed;
    speed_control: process control.speed;
end basic.speed_control;

--
process implementation control.speed
  subcomponents
    scale_speed_data: thread read_data.speed;
    speed_control_laws: thread control_laws.speed;
end control.speed;

thread implementation read_data.speed
end read_data.speed;

thread implementation control_laws.speed
end control_laws.speed;
```

Each component category allows certain other component categories to be subcomponents. For example, you cannot place a thread within a system unless that thread is contained in a process or within a thread group that itself is contained within a process. This restriction is discussed for each category in the Constraints subsection of the respective chapter.

You can partially declare subcomponents by omitting the classifier reference or by simply specifying a component type. For example, consider the system implementation *basic.directional_control* shown in Listing 9-3. In the declaration only the type references are included for the *GPS* and *rudder* subcomponents and no classifier reference is used for the *directional_control* process subcomponent. The result is a partially complete system architecture model, in which some details have not been provided yet. You can still generate instance models from partially complete models and perform some analyses (e.g., you may be able to analyze resource budgets of processes). For other analyses, such as scheduling analysis, the model must include both processors and threads.

Listing 9-3: *Partially Specified Subcomponents*

```
system implementation basic.directional_control
  subcomponents
    GPS: device Basic_GPS;
    rudder: device actuator;
    directional_control: process;
end basic.directional_control;
```

9.1.3 Declaring Subcomponents as Arrays

A subcomponent can be defined as an array consisting of multiple instances of a single component classifier. These can be single or multi-dimensional arrays that are declared by including an array dimension within a subcomponent declaration. An array dimension consists of a size entry that is included before any optional property associations as shown in the box that follows.

```
name : <component category> [component reference] ( [size] )+
       [{properties}] [(in modes)];
```

The size entry defines the number of dimensions and the size of each dimension using square brackets. The plus sign superscript on [size] indicates one or more bracketed entries. Each bracket represents a dimension of the array and an integer, integer property value, or integer constant within the bracket defines the number of elements (the size of the dimension). For example, a two-dimensional array of size two by three is declared as [2] [3]. Arrays may be specified with dimensions but without a size value. The size value is then provided for all dimensions of an array in a refinement declaration. The size value can be an integer, a property, or a property constant. By using a property or property constant you can parameterize the array size.

You can declare arrays at any level of a component hierarchy and for any component category. Examples of declaring array subcomponents are shown in Listing 9-4. The subcomponent *sensor_network* is a

4 by 4 array of instances of the device type *radar_sensor*. The process *data_filters* is a one dimensional array of size *r_level*, where *r_level* is an aadlinteger property defined in the property set *ex_prop_set* and assigned the value 3.

Listing 9-4: *Array Subcomponents*

```
subcomponents
sensor_network: device radar_sensor [4] [4];
data_filters: process filters.basic [ex_prop_set::r_level];

*****
ex_prop_set::r_level => 3;
```

You are able to declare subcomponent arrays of components that themselves contain subcomponent arrays. For example, you can declare a system called *MemoryBoard* that contains a single-dimensional array of memory components, and then declare a system called *MemoryBox* that contains a single-dimensional array of memory boards. The resulting instance model effectively has a two-dimensional array of memory instances.

9.2 Modes

A component can have modes and mode transitions. We can use modes to represent various operational states of a system or component. A mode can represent a configuration of active subcomponents and connections. These configurations can represent different operational configurations and different fault tolerant configurations. For example, during taxiing of an aircraft its autopilot is not engaged. Similarly, an aircraft may have a dual redundant flight guidance subsystem that operates nominally with both instances active, but continues to operate with one instance when the other instance fails. Mode transitions determine when the system is reconfigured dynamically to a new configuration.

Components can have property values that differ from mode to mode. For example, a thread can have different modes to represent an algorithm executing with different levels of precision and different execution times, or a processor can have different modes with different cycle times.

Modes can be used within threads or subprograms to specify different call sequences. This allows you to model the behavior of application code. The mode concept has some limitations in that role since it is primarily intended for architectural modeling. Therefore, you can make use of the Behavior Annex standard [BAnnex] or a state-based modeling language to represent the functional behavior of application components.

9.2.1 Declaring Modes and Mode Transitions

Modes are states within a state machine abstraction. They are declared in the modes section of a component type or implementation.

A mode declaration involves naming the mode, optionally identifying the mode as the initial mode, and declaring properties as desired. In the declaration, the reserved word **mode** must be included as shown in the template that follows.

```
name : [initial] mode [{properties}];
```

A mode transition declaration consists of a name (that is optional), source mode, triggers, and destination mode for the transition. All transitions involve a single source and a single destination mode. Multiple source or destination modes are not permitted, but you can declare several transitions originating in a particular mode or being the destination of a particular mode.

The template for a mode transition declaration is shown in the box that follows.

```
[ name :] (source mode)  -[ (triggers) ]-> (destination mode) ;
```

Note that the brackets shown with the transition arrow are part of the mode transition symbols, not a designation of an optional entry. Rather, a mode transition symbol consists of two parts −[and]-> that enclose the triggers for the transition. The triggers for a transition are port names (event, event data, or data), a **self**.*source_name* declaration, or a **processor**.*source_name* declaration. The named ports can be those of the component itself to represent external events triggering mode transitions, or ports of subcomponents to represent events propagating up the component hierarchy and being handled by this component. You can also name events that are raised by the component itself such as fault event, or by the processor on which the application executes.

The arrival of events, data, or event data through a named port is the trigger for a transition. If the mode transition names more than one trigger, then the arrival of events, data, or event data on any of them results in a mode transition. In other words, the mode transition trigger condition is a logical *or* condition. The default *or* condition can be refined through an annex subclause into a more complex condition. Such conditions can be evaluated by application software, such as a health monitor, which then raises the mode transition trigger event.

9.2.2 Declaring Modal Component Types and Implementations

Modes can be declared within component types or implementations. When declared in a type, modes apply to all implementations of that type. There are two cases of mode declarations within component types. In the first case, modes are declared with mode transition declarations in a **modes** section. In this case, the mode state machine performs mode transitions according to the transition trigger specification. In the second case, modes are declared as inherited in a **requires modes** section. In this case, the component inherits the mode state machine from the enclosing component and operates in different modes according to the mode transition of the inherited mode state machine. Inherited modes are described in more detail in Section 9.2.4.

Mode transitions may be triggered externally through ports in a component type. In this case, the mode transitions can be specified in the component type. A mode transition may be triggered from within the component. In this situation, component implementations can add mode transitions that refer to the event ports of subcomponents. However, component implementations cannot add modes, if modes are declared in the component type.

Examples of mode declarations are shown in Listing 9-5 that presents textual and graphical representations of modes and transitions for a simplified controller thread type *control_modal*. In this example, mode transitions are triggered by external events that enter through event ports. There are two modes, *idle* and *controlling*, and three event ports in this example. The *idle* mode is the initial mode. The event brought into the thread by the event port *cc_engage* results in a mode transition from the *idle* mode to the *controlling* mode (the thread configuration that provides the functionality to maintain a set speed). The event carried through the event port *cc_resume1* also results in a switch to the *controlling* mode. The event port *cc_brake* results in an exit of the *controlling* mode and transition into the *idle* mode.

Listing 9-5: *Example Modes Within a Type*

```
thread control_modal
features
cc_engage : in event port;
cc_resume : in event port;
cc_brake: in event port;
modes
idle : initial mode;
controlling : mode;
idle -[ cc_engage, cc_resume ]-> controlling;
controlling -[ cc_brake]-> idle;
end control_modal;
```

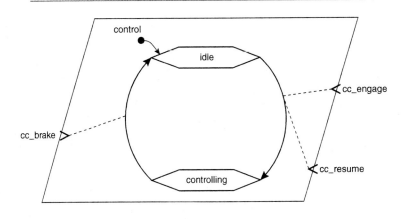

Listing 9-6 shows type and implementation declarations for a system with the same modes and transitions shown in Listing 9-5. However, the modes are declared in the implementation *control_b.cc_control* rather than in the type *control_b*. The lower portion of Listing 9-6 contains a graphical representation of the mode transitions for the implementation. Since no modes are declared in the type, other implementations of *control_b* could define alternative modes from those of *control_b. cc_control*.

Listing 9-6: *Sample Graphical and Textual Specifications for Modes*

```
thread control_b
features
cc_engage : in event port;
cc_resume : in event port;
```

```
cc_brake: in event port;
end control_b;

thread implementation control_b.cc_control
modes
idle : initial mode;
controlling : mode;
idle -[ cc_engage, cc_resume ]-> controlling;
controlling -[ cc_brake]-> idle;
end control_b.cc_control;
```

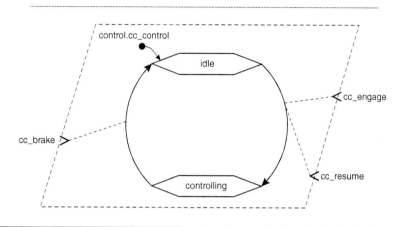

9.2.3 Using Modes for Alternative Component Configurations

Modes can be used to define alternative configurations of subcomponents and connections within an implementation. This can be done when the modes and their transitions are declared in the implementation or in its type. You indicate which subcomponents and connections are part of a mode by adorning subcomponents and connections with an **in modes** statement. Within threads and subprograms, you can do the same for call sequences. In addition, you can declare flow specifications, flow implementations, end-to-end flows, and property associations to be mode-specific with an **in modes** statement.

The template for an **in modes** statement is shown here.

in modes (mode names)

The mode names entry is a single mode name, or a list of mode names separated by commas. Note that the parentheses are required around mode names entries. The mode name must refer to a mode declared within the component or as inherited by the component through the

requires modes section. You can also specify that a connection is only active during a mode transition. In this case, you refer to the mode transition name in the **in modes** statement of the connection declaration. This capability is useful if you want to model that data is transferred from a component that is deactivated to a component that is activated during a mode transition.

The example in Listing 9-7 shows a multimode *process control_algorithms.impl*. In the textual specification for the process *control_algorithms.impl*, the **modes** section defines the two operational *modes* of *ground* and *flight* and the transitions between them. The transitions are triggered by out event ports from the thread *controller* that is a subcomponent of the process implementation *control_algorithms.impl*. The subcomponent thread *controller* is active in all modes, whereas the thread *flight_algorithms* is active only in the *flight* mode and the thread *ground_algorithms* is active in the *ground* mode. Similarly, the connections to and from the data ports on the thread *flight_algorithms* are active only during *flight* mode and the connections to and from the data ports on the thread *ground_algorithms* are active in the *ground* mode.

In the upper right portion of the figure in Listing 9-7, a graphic shows the *modes* and their transitions that are triggered by the events from the *controller* thread. In that figure, the *flight* mode configuration is shown in black and the *ground* mode is shown in gray. This distinction illustrates that the *ground_algorithms* thread and its connections are not part of the *flight mode*.

Listing 9-7: *Modes Example*

```
process control_algorithms
features
status_data: in data port;
aircraft_data: in data port;
command: out data port;
end   control_algorithms;
--
process implementation control_algorithms.impl
subcomponents
controller: thread controller;
ground_algorithms: thread ground_algorithms in modes (ground);
flight_algorithms: thread flight_algorithms in modes (flight);
connections
C1: port aircraft_data -> ground_algorithms.aircraft_data
    in modes (ground);
```

```
C2: port aircraft_data -> flight_algorithms.aircraft_data
    in modes (flight);
C3: port ground_algorithms.command_data -> command
    in modes (ground, g_to_f);
C4: port flight_algorithms.command_data -> command
    in modes (flight);
modes
ground: initial mode;
flight: mode;
g_to_f: ground -[controller.switch_to_flight]-> flight;
f_to_g: flight -[controller.switch_to_ground]-> ground;
end control_algorithms.impl;
--
thread controller
features
status_data: in data port;
switch_to_ground: out event port;
switch_to_flight: out event port;
end controller;
--
thread ground_algorithms
features
aircraft_data: in data port;
command_data: out data port;
end ground_algorithms;
--
thread flight_algorithms
features
aircraft_data: in data port;
command_data: out data port;
end flight_algorithms;
```

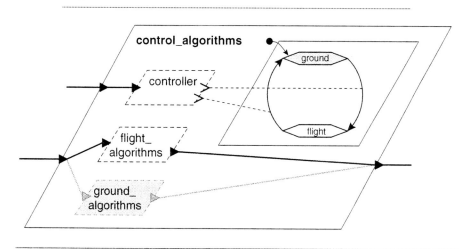

9.2.4 Inheriting Modes

As mentioned in Section 9.2.2, components can declare their modes to be "inherited" from their enclosing component by declaring the modes using a **requires modes** statement. When you declare an instance of such a component as a subcomponent, you must specify a mapping of the enclosing modes to all the modes of the subcomponent. You do this through the **in modes** statement of the subcomponent declaration. For each mode, you specify *<enclosing mode>* **=>** *<subcomponent mode>*. The names of the enclosing and subcomponent modes may be the same or different. Multiple modes of an enclosing component may be mapped to a single subcomponent mode.

This capability allows a process to have modes with subcomponent threads that all respond to the same mode state machine. As a simple example, consider the AADL text shown in Listing 9-8, in which there is a thread subcomponent *control* for the process implementation *autopilot.modal_thread*. This thread has modal behavior, where the thread's execution time is mode dependent. The modes for the subcomponent thread are declared in the **requires modes** statement in the subcomponent's classifier type declaration *controller*. The modes for the containing process implementation *autopilot.modal_thread* are defined in its process type declaration *autopilot*. In declaring the thread subcomponent *control*, the mode names are mapped from the containing process to the thread subcomponent (*ground => ground, flight => robust*).

An example of an implementation of such a modal behavior utilizes a state variable. This state variable in the source code is accessed by all subcomponent threads of the process. One subcomponent thread sets the variable thereby managing the state changes. The other threads read this variable and respond appropriately (e.g., executing alternative algorithms with different execution times). By modeling this implementation with inherited modes, we are able to be explicit about mode-specific component properties. If we had modeled the state variable as a data component with shared access by threads through requires data access features, we would have lost the fact that the state variable represents mode states.

Listing 9-8: *An Example of Inheriting Modes*

```
process autopilot
features
switch_to_flight: in event port;
switch_to_ground: in event port;
status_data: in data port;
```

```
aircraft_data: in data port;
command: out data port;
modes
ground: initial mode;
flight: mode;
ground -[switch_to_flight]-> flight;
flight -[switch_to_ground]-> ground;
end autopilot;
--
process implementation autopilot.modal_thread
subcomponents
control: thread controller.modal
    in modes (ground => ground, flight => robust);
status_manager: thread status_mgr.basic;
end autopilot.modal_thread;

thread controller
features
switch_to_flight: in event port;
switch_to_ground: in event port;
status_data: in data port;
aircraft_data: in data port;
command: out data port;
requires modes
ground: initial mode;
robust: mode;
end   controller;

thread implementation controller.modal
properties
Compute_Execution_Time => ( 2 ms..5ms in modes (ground),
                            3 ms..7ms in modes (robust));
end controller.modal;
```

9.2.5 Mode-Specific Properties

Mode-specific property associations can be used to define mode-dependent property values. The property values can only be dependent on the mode of the component that the property value is associated with. For example, in Listing 9-8 the thread *controller.modal* has two property values, one for each of the two modes of the thread. In the example, the values are dependent on the two modes of the thread (defined as Ground and Robust in Listing 9-8, not the modes of the process that the thread subcomponent is declared in).

For example, consider the partial specification in Listing 9-9 that has a modified version of the process implementation for *control_algorithms.impl* shown in Listing 9-7. In this example, the modes for the process are defined in the process type *control_algorithms*. The mode transitions do not have an identifier and are triggered by internal events (the **self** declarations). The thread subcomponent *controller* in the process implementation *control_algorithms.impl* has a different execution time for the *ground* mode than for the *flight* mode. Note that multiple property associations are used to adorn a single declaration for that subcomponent declaration.

Listing 9-9: *Mode-Specific Component Property Associations*

```
process control_algorithms
features
status_data: in data port;
aircraft_data: in data port;
command: out data port;
modes
ground: initial mode;
flight: mode;
ground -[self.switch_to_flight]-> flight;
flight -[ self.switch_to_ground]-> ground;
end   control_algorithms;
--
process implementation control_algorithms.impl
subcomponents
controller: thread controller {Compute_Execution_Time =>
                ( 2 ms..5ms in modes (ground),
                  3 ms..7ms in modes (robust) ) };
ground_algorithms: thread ground_algorithms in modes (ground);
flight_algorithms: thread flight_algorithms in modes (flight);
--
end control_algorithms.impl;
```

9.2.6 Modal Configurations of Call Sequences

Alternative *call* sequences can be specified using modes. The example in Listing 9-10 shows a *monitor* thread that checks software and hardware and reports anomalies. The thread employs a sequence of calls to subprograms when the thread is in the *nominal* mode. When an error is detected, an *error_condition* is signaled through the event port *error_event*. This signal results in a mode switch and changes the subprogram calls sequence of the thread.

Listing 9-10: *Mode-Dependent Call Sequences*

```
thread monitor
features
error_event: in event port;
```

```
repaired: in event port;
end monitor;
--
thread implementation monitor.impl
calls
        nominal_sequence: {
                call_cksw: subprogram check_sw;
                call_ckhw: subprogram check_hw;
                call_report: subprogram report;
                } in modes (nominal);
    error_sequence: {
                call_alarm: subprogram alarm;
                call_diag: subprogram diagnose;
                callreport: subprogram report;
                } in modes (error_condition);
modes
nominal: initial mode;
error_condition: mode;
nominal -[error_event]-> error_condition;
error_condition -[repaired]-> nominal;
end monitor.impl;
```

Chapter 10

Component Interactions

You define component interactions by establishing relationships between the components using connections, calls, and bindings, as summarized in Table 10-1. You use connections to define interactions of components through externally visible features. There are five sets of features: ports, access (data, subprograms, and buses), parameters, feature groups, and abstract features. You use calls to identify the

Table 10-1: *Component Interactions*

Connections	
port	Relationships between ports as well as relationships between ports and data components that enable the directional exchange of data and events
access	Relationships that enable components to access data, bus, subprogram components
parameter	Relationships among data elements associated with subprogram calls
feature group	Single unit relationships between feature groups outside of components
abstract feature	Relationship between placeholder features

continues

Table 10-1: *Component Interactions (continued)*

Calls	
subprogram	Relationships within component implementations that enable synchronous call/return access to subprograms
Binding	
binding	Relationships that define the mapping of software components, connections, and subprogram calls to hardware components

subprogram being called by the calling components (threads and subprograms). You use binding property associations to declare mappings of software to hardware elements and mappings of calls to remote procedures.

10.1 Ports and Connections

A port is an interface for the directional transfer of data, events, or both (data and events) into or out of a component. Port connections are pathways for such directional transfers between components

10.1.1 Declaring Ports

Ports are named features in component type declarations. The template for a port declaration is shown in the following box.

```
name : <direction> <port type> [data identifier] [{properties}];
```

The direction of the information flow through the port is defined using port direction reserved words: **in, out,** or **in out**. The direction **in** represents a component's input, while **out** represents a component's output. An **in out** direction indicates that a port is used for both input and output. Incoming and outgoing connections can be made to the same component or different components.

The nature of a port, whether it involves queued data transfer, queued event transfer, or non-queued data transfer, is defined by the port type reserved words: **event data port, event port,** or **data port.** The optional data identifier that can be included for data and event data ports defines the nature of the data that can pass through the port. This may be the name of a data type or a data implementation. The optional property associations describe characteristics of the port such as the variable name associated with the port.

Event data ports are ports for asynchronously sending and receiving data such that the data may be queued if the receiving component is busy. They represent message ports. Event ports are event data ports without data content. They represent discrete events such as a change in switch position or a processor clock interrupt. Data ports are event data ports that retain only the most recent arrival. They represent sampling ports (i.e., ports without a queue) for communicating state information, such as signal streams that are sampled and processed periodically.

Listing 10-1 includes five example port declarations for a process. The data type and the name of the source variable for the port are included in the declaration for the in data port *speed*. Notice that the source variable name need not be identical to the port name or data type in the AADL model. All of the optional elements of the port declaration are omitted in the declaration of the out event data port *Error_Signal*.

Listing 10-1: *Example Port Declarations*

```
process control
features
speed: in data port raw_speed {Source_Name => "RawSpeed";};
disable: in event port {Required_Connection => false;};
set_speed: in event data port raw_speed.setpt;
throttle_cmd: out data port command_data;
Error_Signal: out event data port;
end control;
```

The graphical representations and their corresponding declarations for data ports, event ports, and event data ports are summarized in Table 10-2.

In addition to user-defined ports, there are three predeclared ports for threads: *Dispatch*, *Complete*, and *Error*. While these are available for use directly in connections to and from the thread, it is not necessary to declare them explicitly. Dispatch is an in event port that can be used to initiate the execution of the thread. Complete is an out event port that signals the completion of a thread. Error is an out event data port that can be used to report thread execution errors.

Table 10-2: *Ports as Interfaces*

Graphical Representation	Description
Data Ports in out ▶ ▶ in out ◆	**Data ports** are interfaces for state data transmission among components without queuing. Within source code, they are represented by typed variables. The structure of the variable/array is defined by a data type [data classifier] associated with the port.
Event Ports in out ❯ ❯ in out ◇	**Event ports** are interfaces for the communication of events raised by subprograms, threads, processors, or devices that may be queued. Examples of event port use include: triggers for the dispatch of an aperiodic thread, initiators of mode switches, and alarm communications. Events such as alarms may be queued at the recipient, and the recipient may process the queue content. Event ports are represented by variables within source code that are associated with runtime service calls.
Event Data Ports in out ▶ ▶ in out ◆	**Event data ports** are interfaces for message transmission with queuing. These interfaces enable the queuing of the data associated with an event. An example of event data port use is modeling message communication with queuing of messages at the recipient. Message arrival may cause dispatch of the recipient and allow the recipient to process one or more messages. These ports are represented by variables in source code that are associated with relevant runtime service calls.
out in in out out out in out out in	Graphically, **in** and **out ports** are identified by the relative position of their icon on the border of the component for which it is a feature. Out port icons are located on the exterior face of a component's boundary and in port icons are located on the interior face of a component's boundary. They can be located on any boundary of the component icon. The in out ports straddle the boundary of the component icon. Representative positioning is shown in the graphic to the left for a system component.

10.1.2 Declaring Port to Port Connections

The template for a textual declaration of a port to port connection is shown in the following box.

```
name : port <source port> <connection symbol> <destination port>
     [{properties}] [in modes and transitions];
```

In a port connection declaration, we give the connection a name and refer to the source port and the destination port. A connection can be directional (→) or bi-directional (↔). For directional connections, the source is on the left of the connection symbol and the destination is on the right. In a bi-directional connection, the information can flow in both directions or is determined by the directions of the ports on either end of the connection. If the ports are **in out** ports, the bi-directional connection indicates the transfer of data in both directions.

The properties for connections include properties for identifying protocols associated with the connection, the quality of service expected from such a protocol, the binding of the connection onto networks (AADL bus) or processors, and connection patterns between component arrays. You can find these properties in the *Deployment_Properties* property set.

The optional "in modes and transitions" entry associates a connection with specific modes or mode transitions. In other words, a connection may be active in one mode and not in another. Similarly, a connection may only be active during a mode transition (e.g., to exchange state information from a thread that becomes inactive to a thread that becomes active).

10.1.3 Using Port to Port Connections

You declare port to port connections between ports of subcomponents within a component implementation. By doing so you indicate that two subcomponents or subcomponents contained inside each of them communicate data and events.

You also declare port to port connections between a port of a subcomponent and a port of the enclosing component. In this case, you indicate that the output of a subcomponent port is passed to an external component via the port of the enclosing component, and that the input to a subcomponent port comes from an external source via the port of the enclosing component. In other words, port to port connections between two threads in different processes must be declared by following the component hierarchy. For example, two threads in different processes can only communicate with each other if their enclosing processes provide ports for that purpose and the ports are appropriately connected. Following the component hierarchy when declaring port to port connections results in a modular system design. It does not create runtime overhead as in the actual system instance as data is sent directly from one thread to another.

Listing 10-2 has an example textual specification and corresponding graphical representation that includes port and port connection declarations. Within component type specifications, appropriate ports declarations are grouped together in the features section. Supporting data type definitions are included at the end of the table. We could have placed these data type declarations in a separate package.

In Listing 10-2, the connection *c_data_transfer* is a delayed connection[1] between the out data port *c_data_out* of the thread input (written as *input.c_data_out*) and the in data port *c_data_in* of the thread *control_plus_output* (written as *control_plus_output.c_data_in*). The connections declaration *brake_in*: **port** *brake -> input.brake_event*; connects the in event port *brake* of process implementation *control.speed_control* to the in event port *brake_event* of the thread subcomponent input. The optional name for the port connection between *control_plus_output.c_cmd_out* and *throttle_cmd* is not included in this example. The predeclared event data port *Error* is used as the source in the connection *error_connection*. This event data port is not declared as a feature of the originating thread, since it is a standard feature of all threads. Note that the *c_data_transfer* connection is between two subcomponents, while the other connections declare a mapping between a subcomponent port and a port of the enclosing component.

Graphically, port connections are solid lines between the ports involved in the connection. These can include adornments that define the timing of the connection. For example, double vertical bars adorn a delayed connection (a data transfer that is phase delayed) between data ports for data transfer between periodic threads. A delayed connection is shown in the graphic of Listing 10-2 for the data port connection between the threads *input* and *controller*. A double arrow indicates an immediate connection (a data transfer that is within the same frame).

Note the connections between ports on a subcomponent and ports on the boundary of the component in which that subcomponent is

---

1. Timing of connection communication is discussed in Section 10.1.5.

contained. The names of the ports involved in these connections can be different. However, these ports must be of the same port type and direction. In addition, for event data and data ports, they must be defined as transferring the same data (e.g., in Listing 10-2 the in data port *speed* of the control process has the same data declaration *raw_speed* as the in data port to which is connected the *speed_in_data* of the thread type *control_in*). This means that their data types must match. By default, the data types must be identical, but other matching rules, such as equivalence and subset, can be specified. You can use equivalence to indicate that two data types that have been defined independently represent the same data. You can use subset to indicate that the recipient data type represents a subset of the record fields of the sender. Note that a port may be the source of multiple connections, as shown for the connections *brake_in1* and *brake_in2* within the process implementation *control.speed_control* that originate at the event port *brake*.

Communication between data ports on the same processor or processors with shared memory can be mapped into static data variables in the source code unless double buffering is required. Double buffering may be required because the sending and receiving threads execute concurrently or they execute with different periods.

Listing 10-2: *Sample Declarations of Data, Event, and Event Data Ports*

```
process control
features
speed: in data port raw_speed;
brake: in event port;
set_speed: in event data port raw_set_speed;
throttle_cmd: out data port command_data;
Error_Signal: out event data port;
end control;
thread control_in
features
speed_in_data: in data port raw_speed;
brake_event: in event port;
set_speed_edata: in event data port raw_set_speed;
c_data_out: out data port processed_data;
end control_in;

thread control_out
features
brake_event: in event port;
c_data_in: in data port processed_data;
c_cmd_out: out data port command_data;
end control_out;
```

continues

```
process implementation control.speed_control
subcomponents
input: thread control_in.input_processing_01;
controller: thread control_out.output_processing_01;
connections
speed_in: port speed -> input.speed_in_data;
brake_in1: port brake -> input.brake_event;
brake_in2: port brake -> controller.brake_event;
set_speed_in: port set_speed -> input.set_speed_edata;
c_data_transfer: port input.c_data_out -> controller.c_data_in
    {Timing => Delayed;};
port controller.c_cmd_out -> throttle_cmd;
error_connection: port input.Error -> Error_Signal;
end control.speed_control;

thread implementation control_in.input_processing_01
end control_in.input_processing_01;

thread implementation control_out.output_processing_01
end control_out.output_processing_01;
data raw_speed
end raw_speed;
data raw_set_speed
end raw_set_speed;
data command_data
end command_data;
data processed_data
end processed_data;
```

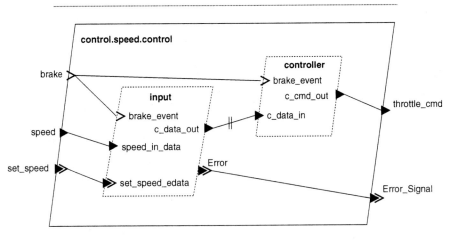

You can use a port to port communication between threads (i.e., two active software components that may reside in the same or different processes and execute on the same or different processors). You can also connect a thread port to a device port to model the logical interaction between a physical component and the embedded application or between two devices. Finally, you can model the interaction between an application thread and its runtime system by a port connection to a processor port. For example, you can represent exceptions reported by the runtime system as event ports or event data ports. In a port to port connection declaration you refer to a processor port by "**processor** . <portname>", where **processor** identifies the processor the thread executes on.

10.1.4 Constraints on Port to Port Connections

The type of the ports that are connected need not be identical. For example, connections can be made between data ports and event data ports and from event data ports to event ports. The type of the destination port determines how the data or event is transferred and received. An incoming data port can sample the output from a data port or event data port. In this case, it samples a stream of state values or a stream of messages, possibly skipping some sent values or messages or sampling them multiple times. You will use sampling data ports to model sampled processing in control systems, or sampling of message streams to determine whether messages are being sent within a time interval. An event data port can receive and queue output from an event data port and from a data port. In this case, every sent message or state value of the sender is queued to be processed without loss. The output from a data port can be sent to both a data port and an event data port. You may use this to model a control system sampling sensor data and a logging capability recording every sensor reading. An event port can observe and record as a queued event, the arrival of data from a data port, a message from an event data port, or an event from event port. You may use a connection from a data port or event data port to an event port to model a system health monitor that is interested in determining whether the sender is alive. Permitted connections are summarized in Table 10-3.

Table 10-3: *Inter-Port Connections*

From \ To	Data	Event Data	Event
data	Yes	Yes	Yes
event data	Yes	Yes	Yes
event			Yes

Port to port connections require that the direction of the source port matches the direction of the destination port. For connections between ports of two subcomponents this means that the source port must be an outgoing (out or in out) port, while the destination port must be an incoming (in or in out) port. For connections up the hierarchy from a subcomponent port to a port of the enclosing component, both ports must be outgoing, while for connections down the hierarchy from an enclosing component to a subcomponent both ports must be incoming.

There are restrictions on the topology of port connections. An out data port can be connected to multiple in data ports—a "fan-out" of data to multiple destinations. In this case, each destination port receives an instance of the transmitted data. However, since data ports do not support queuing, an in data port is restricted to a single incoming connection (i.e., cannot have multiple connections from different sources). Since data ports hold a single value, without queuing, outputs from multiple sources would overwrite one another. You should be aware that if you have a system with multiple modes, a thread with an in data port can have a different single incoming connection in each mode.

In contrast, since they support queuing, event and event data ports can have multiple input (fan-in) connections as well as multiple output (fan-out) connections. Multiple inputs at an event or event data port enable the specification of the sequencing as well as the queuing of events.

Data and event data ports include data component classifier references to represent the data type of the data communicated through the port. AADL requires that the data type of the source port matches the data type of the destination port. For example, the connection from the out data port of the thread *read* to the in data port of the thread *scale* in Listing 10-2 includes the same data type declaration for each of the ports. By default, the data types have to be identical. However, you can specify other matching rules with the *Classifier_Matching_Rule* property, such as the destination accepting a subset of the source data, or the protocol to be used in the connection supports conversion between the two data types. The reference to a data component classifier is optional to support creating partial specifications[2]. Therefore, it is acceptable for one end of a connection not to have a data type declared while the other end does. Similarly, one end of a connection can have just a data

---

2. Although these omissions are permitted, within the OSATE environment a warning message will be generated to ensure that a user is aware of this situation.

component type, while the other end has a data implementation with the same type.

Typically, you want to perform additional consistency checks on port connections that involve data exchange. For example, you may want to ensure that the base type used by the sender to represent the application data type matches that of the receiver (e.g., that both use 16-bit signed integer to represent temperature). Similarly, you may want to check that both application data types use the same measurement unit (e.g., Celsius). The AADL Data Modeling Annex standard defines a set of properties to record such information.

A port with **in out** direction represents a single interface into and out of a component, in effect both an in port and an out port. You can declare a single bidirectional port to port connection between two in out ports. In this case, the two components communicate in both directions. You can also declare multiple directional connections to an *in out* port as shown in Listing 10-3, where the in out port *error_data* on the process *monitor* receives data from the out data port *error_data* of the process *input* and outputs a value to the in data port *error_response* of process *input*. The out data port *error_data* of process input is connected to two receiving ports, one on the process *master* and the other on the process *monitor* via an immediate connection. The connection from the in out data port *error_data* on the process *monitor* to the *error_response* in data port on the process *input* is a delayed connection[3]. The graphic in the lower portion shows the topology for these connections.

As with unidirectional data ports, an in out data port can be mapped to a single static data variable within application source code. Since an in out port maps to a single static variable, an existing incoming value of a port will be overwritten by the source code when an output value is written to that port.

Listing 10-3: *Connections with in out Ports*

```
system basic
end basic;

system implementation basic.monitor
  subcomponents
    input: process input;
```

continues

---

3. We are assuming that there is a single periodic thread within the process *input* that sends data through the data port *error_data* and receives data through the data port *error_response*.

```
    monitor: process monitor;
    master: process master;
  connections
    C1: port input.error_data -> master.error_data
    {Timing => Immediate;};
  C2: port input.error_data -> monitor.error_data
    {Timing => Immediate;};
  C3: port monitor.error_data -> input.error_response
    {Timing => Delayed;};
end basic.monitor;

process input
features
error_response: in data port;
error_data: out data port;
end input;

process monitor
features
error_data: in out data port;
end monitor;

process master
features
error_data: in data port;
end master;
```

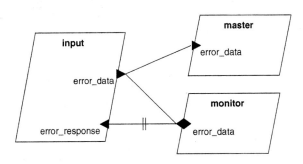

10.1.5 Port Communication Timing

Port to port connections define communication between active components of a system (e.g., between two threads, a thread and a device, or
two devices). The port type and the dispatch protocol of the thread or
device determine the timing of the communication. Periodic dispatch
protocols allow for sampling of input, while in other cases the arrival of
events or data can trigger the dispatch as discussed in Section 6.1. In
the case of port communication between two periodic threads/devices

through data ports, AADL supports sampled processing with up- and down-sampling as well as deterministic sampling. The latter ensures that latency jitter is minimized—an important factor for control systems whose stability is sensitive to such jitter.

For outgoing event and event data ports the transfer occurs at the time it is initiated by the sender through a *Send_Output* service call in the source code. This time can be specified through the *Output_Time* property on the outgoing port. Once the sent output is received by a destination event or event data port, it is placed in its input queue. The arrival of the message may trigger a dispatch of the recipient thread, if it is listed as a dispatch trigger port for the thread. By default, the one message in the input queue is "frozen" at thread dispatch (i.e., it is removed from the queue and made available to the source code in a port variable). The content of this port variable is guaranteed not to change during the execution of the thread, even though new messages may arrive. This means any messages or events arriving after the dispatch are queued, but not available to the recipient until the next dispatch. You can indicate through a *Dequeue_Protocol* property that the whole queue content should be frozen and made available to the thread source code for processing in the current dispatch. For example, this is useful when you have a health monitor that periodically wants to examine alarm events or messages. You can also indicate that the input should be "frozen" during the execution of the thread by using the *Input_time* property on the incoming port. In the source code, it is expressed by an explicit *Receive_Input* service call. The number of queued items that have been frozen can be retrieved by a *Get_Count* service call. If the queue is empty at the time of a freeze, the count will be zero.

In the case of an event port connection, the source may be an internal event. This is declared by using **self**.*eventname* as the source. The identifier *eventname* is a specific source internal to the component containing the connection. For example, in the case of a thread the internal event may originate as an exception from the source code of the thread or from the execution platform on which the thread is executing. If the containing source is a process or thread group, the internal event can represent an event related to a fault of this component; the fault condition may be modeled with the Error Model Annex [EAnnex].

For outgoing data ports the data transfer is by default initiated at completion time. You can indicate a send during the execution through the *Output_Time* property. In that case, the transfer is initiated by an explicit *Send_Output* call in the source code.

For incoming data ports the content is "frozen" the same way as for event data ports. In other words, by default it is frozen at dispatch time, or otherwise at a time specified by the *Input_Time* property. Only the most recent value is available to the source code. If no new data has been received since the last dispatch, the old value is still available, but marked as not fresh. The source code can determine the freshness through an *Updated* service call.

10.1.6 Sampled Processing of Data Streams

In this scenario, the sender and receiver can be threads or devices. For port connections with a periodic recipient with a data port, the connection represents sampling communication. This means that the receiver samples a data stream at a given rate. The receiver samples independent of the rate at which the sender executes and sends data. If the sampling rate differs from the rate at which the data is sent, over- or undersampling may occur. You can typically find independent sampling in systems with processors and devices interconnected by buses or networks and operating on independent clocks.

In a sampling data port connection the receiver reads the output of a source at the receiver's dispatch time. This data transfer occurs independently from the execution characteristics of the sending thread or device. An offset time can be defined by assigning a value to the *Input_Time* property for the receiving data port. This property specifies the amount of execution time after dispatch before the input is read.

The rate at which data arrives may be the same, higher, or lower than the sampling rate and it may be constant (sent by a periodic sender) or variable (sent by a sender whose execution is triggered by random events). As a result, the receiver may sample the same input multiple times (over-sampling) or the receiver may not sample all elements of the data stream (undersampling). In the case of a variable data stream, a sampling thread will turn such a data stream into an output data stream with a periodic rate.

Since sampling communication is independent of the execution of the source component, the data stream is sampled non-deterministically. Consider two threads that are executing concurrently on different processors. While each thread will execute periodically at 20Hz, the relative start times for execution of the threads can vary. These variations will result in non-deterministic sampling, as illustrated in Figure 10-1. In this example, within the first 50ms (20Hz) frame of the sequence, thread *control* receives the output of the initial dispatch of *read_data* (the value

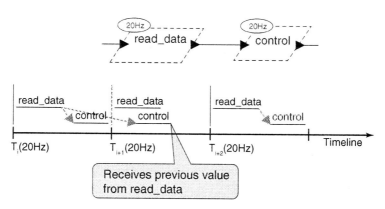

Figure 10-1: *Non-Determinism in Sampling Connections*

output in the same frame). However, in the second 50ms (20Hz) frame the thread *control* again receives the output of the initial dispatch of *read_ data* (the result from the previous frame). This results because the second execution of *read_data* is not complete and its output is not available when the thread *control* executes. The third dispatch of *control* may receive the output from the third dispatch of *read_data*, as shown. Unlike the deterministic semantics of immediate and delayed connections, the semantics of a sampling connection does not constrain the order of execution of communicating threads or devices and non-determinism may occur in the communication.

10.1.7 Deterministic Sampling

If both the sender and the receiver thread are periodic, you can specify that their data transfer is coordinated, such that *deterministic sampling* occurs. You can indicate that the transfer should be *immediate*, having the effect of guaranteed mid-frame communication, or as *delayed*, indicating phase-delayed communication, using the *Timing* property on the connection. Deterministic sampling limits the amount of jitter in data latency. Jitter can affect the stability of control systems. You can find phase-delayed communication in systems that synchronize around a periodic bus, such as a CAN bus or a MIL-STD 1553 bus. Mid-frame communication is typically used within a processor, but can be supported across tightly coupled processors by coordinating the completion and dispatch of the sender and the receiver.

For immediate data port connections between periodic threads, data transmission is initiated when the source thread completes and enters the suspended state. The value delivered to the in data port of a receiving thread is the value produced by the sending thread at its completion. For an immediate connection to occur, the threads must share a common (simultaneous) dispatch. However, the receiving thread's execution is postponed until the sending thread has completed its execution. This aspect can be seen in Figure 10-2, where the immediate connection specifies that the thread *control* must execute after the thread *read_data*, within every 50ms period. In addition, the value received by the thread *control* is the value output by the most recent execution of the thread *read_data*. The ovals in Figure 10-2 labeled with (20Hz) are representations of the thread's execution rate and establish a period of 50ms for the thread that it adorns.

For the graphical timelines in Figure 10-2 and Figure 10-3, a horizontal bar above the timeline that is labeled with a thread name represents the execution time of that thread. The left edge represents the start and the right edge represents the termination of the thread's execution. A solid or segmented arrow between thread execution bars represents a data transfer between threads. A segmented arrow represents a delayed (e.g., Figure 10-3) or a repeat transfer (e.g., Figure 10-4).

For a delayed data port connection between periodic threads, the value from the sending thread is made available at its deadline to the receiving thread at its next dispatch that is the same or follows the deadline[4]. For delayed port connections, the communicating threads

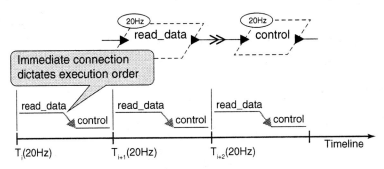

Figure 10-2: *An Immediate Connection*

---

4. If the deadline and the dispatch are at the same time the transfer may be initiated just before the deadline to accommodate for the transfer overhead.

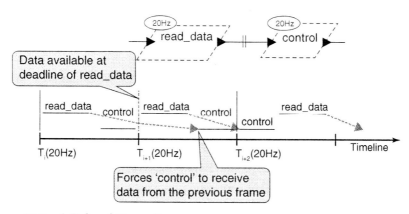

Figure 10-3: *A Delayed Connection*

do not need to share a common dispatch. In this case, the data available to a receiving thread is that value produced at the most recent deadline of the sending thread. If the deadline of the sending thread and the dispatch of the receiving thread occur simultaneously, the transmission occurs at that instant. The impact of a delayed connection can be seen in Figure 10-3, where the thread *control* receives the value produced by the thread *read_data* in the previous 50ms frame. As shown in Figure 10-3, a delayed connection is symbolized graphically by double cross hatching adorning the connection arrow between the ports.

Deterministic oversampling is shown in Figures 10-4 and 10-5. In the case of a delayed connection in Figure 10-4, the value from *read_data*

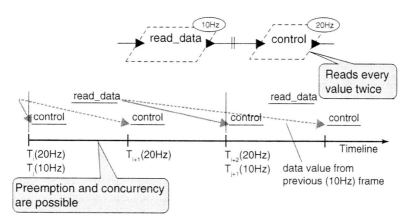

Figure 10-4: *Oversampling with Delayed Connections*

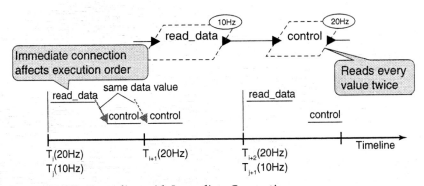

Figure 10-5: *Oversampling with Immediate Connections*

is available at its deadline. It is received by the two executions of *control* whose dispatch coincides with or follows that deadline (e.g., *read_data* may have a pre-period deadline). Thus, the two executions of *control* occurring within an execution frame of *read_data* receive the value produced in the preceding frame of *read_data*.

In contrast, consider the case of immediate connections as shown in Figure 10-5, the values available for two sequential executions of control are the same, the value produced within the 10Hz execution frame of *read_data*. This result is accomplished by delaying the execution of the first control within the frame until the completion of *read_data*. Notice that this can only occur if both *read_data* and an execution of control can successfully complete to meet the deadline within the execution frame of *control*.

Deterministic undersampling is shown in Figure 10-6. For a delayed connection the data provided to an execution of *control* is the value produced by *read_data* that is available at the simultaneous dispatch of

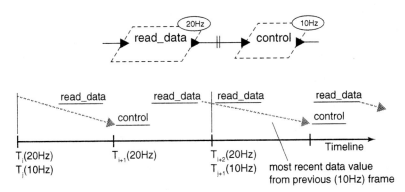

Figure 10-6: *Undersampling with Delayed Connections*

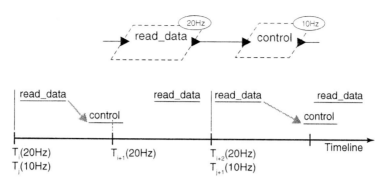

Figure 10-7: *Undersampling with Immediate Connections*

the threads. That value is produced at the most recent *read_data* deadline, which may coincide with the *control* thread's dispatch. In the case of an immediate connection, as shown in Figure 10-7, the value provided to the thread *control* is the value produced by *read_data* at the end of its first execution after the simultaneous dispatch, and the execution of *control* is delayed until *read_data* has completed.

10.1.8 Mixed Port-Based and Shared Data Communication

Sometimes it is necessary for you to combine a subsystem using port-based communication with a subsystem that uses shared data communication. AADL supports this by allowing you to declare connections between ports and data components (or data access features representing access to data components). For example, you can declare a connection from a data component to a data port on a thread. In this case, the content of the data component is read at the time of dispatch of the thread and made available to the application source code in the port variable. This is different from you modeling the thread with a requires data access feature, where the thread can read or write the data component any time during its execution.

You can declare a port connection from a data port or an event data port to a data subcomponent or a data access feature. In this case, a send operation on the port results in writing the sent data value into the data component. You can also declare a port connection from a data component to a data port or event data port. In this case, whenever the data component is written the data is also transferred to the port. Notice that runtime system implementations may not support monitoring of writes to data components to initiate the send to a port.

Listing 10-4 shows example connections between ports and data components. There are data ports connected to the data component *store_p*. The content of *store_p* is updated by the thread *data_mgt* through the connection *to_store_p*. Content from *store_p* is received by the thread *l_control* via the sampling connection *from_store_p*. Directional control information from the thread *l_control* is stored in the data component *store_d* via the connection *to_store_d* from the event data port *output* to the data component *store_d*. The other connections (*data_in* and *cmd_out*) where sensor data is received by and commands are output from the process are also shown. There must be consistency in the data references for connection of ports and data components. For example, the *output* data port on the thread *data_mgt* and the *input* port on the thread *l_control* both refer to the data implementation *store.processed* and the data component *store_p* is an instance of that implementation. Note that the *Access_Right* property for both data components is *read_write*.

Listing 10-4: *Port to Data Component Connections*

```
process control
features
sensor_data: in data port ;
cmd: out data port ;
end control;

process implementation control.directional
subcomponents
data_mgt: thread manage.basic;
l_control: thread controller.basic;
store_p: data store.processed {Access_Right => read_write;};
store_d: data store.direction {Access_Right => read_write;};
connections
to_store_p: port data_mgt.output -> store_p;
from_store_p: port store_p -> l_control.input;
to_store_d: port l_control.output -> store_d;
data_in: port sensor_data -> data_mgt.input;
cmd_out: port l_control.cmd_out -> cmd;
end control.directional;


--
thread manage
features
input: in data port;
output: out data port store.processed;
end manage;

thread implementation manage.basic
end manage.basic;
```

```
thread controller
features
input: in data port store.processed;
output: out event data port store.direction;
cmd_out: out data port;
end controller;

thread implementation controller.basic
end controller.basic;

data store
end store;

data implementation store.direction
end store.direction;

data implementation store.processed
end store.processed;
```

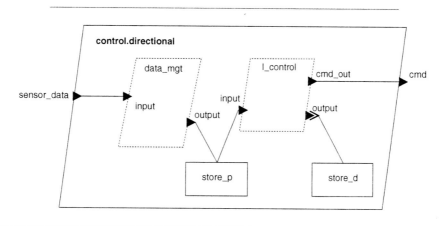

Listing 10-5 extends the example in Listing 10-4 by adding a system that consists of a *nav_processing* and an instance of the process *control.directional* named *control_direction*. The textual specification is a partial listing, including only added or modified portions of the textual specifications from Listing 10-4 and portions of the specification of the system *complete.d_control*. In this example, a provides data access feature *d_data* is added to the process type *control*. It provides access to the data subcomponent *store_d*. A connection is made from the component *store_d* to the access feature *d_data* and from *d_data* to the data port *d_data_in* on the component *nav_processing*.

Listing 10-5: *Port-Data Component Connections*

```
system implementation complete.d_control
subcomponents
control_direction: process control.directional;
nav_processing: process nav;
connections
dc: port control_direction -> nav_processing.d_data_in;
end complete.d_control;


process nav
features
d_data_in: in data port store.direction;
end nav;


process control
features
sensor_data: in data port;
cmd: out data port;
d_data: provides data access store.direction;   -- added
end control;


process implementation control.directional
subcomponents
data_mgt: thread manage.basic;
l_control: thread controller.basic;
store_p: data store.processed {Access_Right => read_write;};
store_d: data store.direction {Access_Right => read_write;};
connections
to_store_p: port data_mgt.output -> store_p;
from_store_p: port store_p -> l_control.input;
to_store_d: port l_control.output -> store_d;
data_in: port sensor_data -> data_mgt.input;
cmd_out: port l_control.cmd_out -> cmd;
d_acc: data access store_d -> d_data; -- added
end control.directional;
```

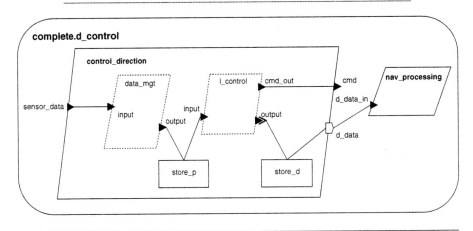

10.1.9 Port and Port Connection Properties

AADL provides a number of predeclared properties for ports. Some properties deal with the fact that the port has a representation in the application source code. For example, *Source_Name* is used to specify the name of the port variable in the source code that is associated with a port. *Source_Data_Size* is inferred from the property on the data component classifier that the data port or event data port references.

Event ports and event data ports can trigger the dispatch of a thread. In that case, you can indicate that the thread executes a different piece of source code for different ports instead of a single entry point for all dispatches by specifying the subprogram to be invoked, execution time, and deadline (if different from that specified for the thread). You can also specify the queue characteristics for event and event data ports, such as size, overflow handling, and dequeuing.

You can use the *Required_Connection* property to indicate whether the component requires a connection on the port or whether it is optional. This allows you to record in the component specification whether its implementation can adapt to an unconnected port (the connection is optional), or whether the component assumes the connection to always be in place. By default, a connection is assumed to be required. You can invoke a consistency check on an AADL model to ensure that this property is satisfied.

AADL provides a number of predeclared properties for port connections. These properties focus on the deployment of the application software (i.e., the mapping of the connection onto the underlying runtime system). You can specify the expected quality of service for the communication protocol and hardware (virtual bus, bus, and processor), such as guaranteed, secure, and ordered delivery. You can also indicate that the connection requires a particular protocol or must occur over a particular network (bound to a particular bus type or bus instance). In some cases, you may want to specify these properties on a port in order to document any assumptions the component makes about communication through a specific port. Finally, you can indicate whether a connection must or must not be co-located with another connection in order to assure redundant communication.

10.1.10 Aggregate Data Communication

You may have a situation where a subsystem (process) is collecting data from multiple sensors or multiple threads produce output that is to be sent to another subsystem. You may have grouped the output from these ports into a feature group at the process level and connected it to the destination subsystem (process). As we discuss in further detail in Section 10.4 a feature group connection represents a collection of individual connections between the ports in the feature group. The threads and devices on the ends of each of the connections communicate independently.

In some cases, you may want to model a system where a subsystem collects the data produced by its components and sends it as an aggregate to another subsystem. You could introduce an extra thread in the subsystem that receives all the outputs, places them in a data record and sends it on a single data port. AADL offers an alternative way of modeling aggregate data communication without having to introduce the extra thread. You define a data component implementation that contains data subcomponents (acting as record field) for each of the data elements of aggregate data record. You then use this data classifier as data type of a single data port in the enclosing process. Within the process implementation, you then connect the output port of each thread or device to a data record field in the process output port. At the destination subsystem, a single thread may receive the aggregate data record or different threads may receive different fields of the data record. As indicated in Section 10.1.4, you can also specify that different threads or subsystems receive different subsets of the aggregate data, modeling the OMG Data Distribution Service [DDS].

The aggregation of data from multiple sources into a single data record is illustrated in Listing 10-6, where the process *controller.impl* contains three threads. Each thread has an out data port that is connected to the aggregate data port *cmd_out* that is a feature of the process *controller.impl*. The port *cmd_out* has a data classifier *agg_data.impl* containing three subcomponents. The connections in the process *controller.impl* are made from the out data port of each thread to a specific data subcomponent of the port *cmd_out*. For example, the connection *cr* is from the port *r_out* on the subcomponent roll to the data subcomponent *cmd_data_r* of the implementation *agg_data.impl*. Note that the data type on the port in the connection must match the data classifiers for the corresponding subcomponent. Access to the data subcomponents of the port *cmd_out* is declared in the type *agg_data*.

Listing 10-6: *Aggregate Data Port Example*

```
data agg_data
end agg_data;
```

```
data implementation agg_data.impl
subcomponents
cmd_data_r: data cmd_data_2;
cmd_data_p: data cmd_data_1;
cmd_data_y: data cmd_data_2;
end agg_data.impl;

data cmd_data_1
end cmd_data_1;

data cmd_data_2
end cmd_data_2;

process controller
features
cmd_out: out data port agg_data.impl;
end controller;

process implementation controller.impl
subcomponents
roll: thread roll;
pitch: thread pitch;
yaw: thread yaw;
connections
cr: port roll.r_out -> cmd_out.cmd_data_r;
cp: port pitch.p_out -> cmd_out.cmd_data_p;
cy: port yaw.y_out -> cmd_out.cmd_data_y;
end controller.impl;
```

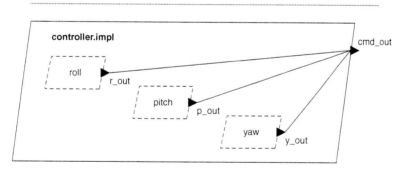

Aggregate data ports are useful if the data from a set of simultaneously dispatched periodic threads must be delivered together. For example in Listing 10-6, the aggregate data port of the enclosing process *controller.impl* ensures that the data from its threads is delivered at the same time. The actual timing of the transmission is determined by the execution of the threads connected to the aggregate data port such that the sending of data is initiated at the completion of the execution of the threads (at the time the last thread completes its execution).

10.2 Data Access and Connections

In your system, multiple threads may operate on a common data area (i.e., they share access to a data component). Similarly, a subprogram may access an external data component by reference. In AADL, you model this through data access features and access connections. You can specify concurrency control on such shared data components through the *Concurrency_Control_Protocol* property on the data component itself.

Data access is declared either as requires access feature indicating that a component needs access to a component, or as provides access feature, indicating that a component allows access to a data component declared within it. The template for an access interface declaration is shown in the following box. The access direction is either provides or requires. The optional component identifier is a data classifier reference that specifies the data type of the data component to be accessed.

```
name : (provides | requires) data access [component identifier]
   [{properties}];
```

You use access connection declarations to specify a path between these data access features and the actual data component to be referenced. Data access connections are similar to port connections with connection name, a source and destination, a directional or bidirectional connection symbol, optional properties, and they can be declared to be active only in certain modes. The template for a data access connection declaration is shown in the following box.

```
name : data access  <source> <connection symbol> <destination>
   [{properties}] [in modes];
```

You can declare the access connection to be bidirectional (↔) or to be directional (→). In the case of a bidirectional access connection read and write access is possible, but may be restricted by the *Access_Right* property on the data component or the data access feature. In the case of a directional access connection the direction indicates reading from a data component (the data component is the source of the access connection) or writing to the data component (the data component is the destination of the access connection) and must be compatible with the *Access_Right* property values if they are specified.

Listing 10-7 presents a system *basic_control.auto_cc* with two sub-systems (processes) that share access to a data component contained in one of them. The thread subcomponent *cc_algorithm* of the process implementation *control.cc_control* is shown to require access to an instance of the data implementation *logs.error_logs*. The *RWLogAccess* data access connection in the process implementation *control.cc_control* connects *comm_error_log* (an instance of *logs.error_logs*) to the requires access feature *error_log_data* on the thread *cc_algorithm*.

Similarly, the thread subcomponent *comm_errors* requires access to the data subcomponent *comm_error_log* (*logs.error_logs*) of the process *cc_error_monitor*. This connection is a remote connection across address spaces, where the process *cc_control* provides access to its data subcomponent. There is a data access connection that extends through the component hierarchy from the data instance *comm_error_log* to the thread *comm_errors*. In addition, there is a data access connection that extends through the component hierarchy from the data instance *comm_error_log* to the thread *comm_errors*, both of which are subcomponents of the process *cc_control*.

Notice the concurrent access to the data subcomponent *comm_error_log* (*logs.error_logs*) in the example. The standard property *Concurrency_Control_Protocol* is used to coordinate the access, as shown in the subcomponent declaration of the data subcomponent *comm_error_log*. The property *Access_Right* is used to define the access protocol of the connection *read_only, write_only,* or *read_write*. This property indicates the desired access. For example if the value is *read-only*, the data component must offer read access and connections to that access feature must be read. If the access is *read-write*, an access connection to the component providing access can be read, write, or both. The property *Access_Time* specifies the period of time at which a shared data component is accessed.

Listing 10-7: *Shared Access Across a System Hierarchy*

```
system basic_control
end basic_control;
system implementation basic_control.auto_cc
subcomponents
cc_control: process control.cc_control;
cc_error_monitor: process monitor.error_monitor;
connections
a_01: data access cc_control.error_log_data -> cc_error_monitor.
error_data_in;
end basic_control.auto_cc;
--
```

continues

```
process control
features
error_log_data: provides data access logs.error_logs
                    { Access_Right => read_only;};
end control;

process implementation control.cc_control
subcomponents
comm_error_log: data logs.error_logs
  {Concurrency_Control_Protocol => Interrupt_Masking;};
cc_algorithm: thread algorithm.cc;
connections
RWLogAccess: data access comm_error_log <->
                        cc_algorithm.error_log_data;
ROLogAccess: data access  comm_error_log -> error_log_data ;
end control.cc_control;

thread algorithm
features
error_log_data: requires data access logs.error_logs
  { Access_Right => read_write;};
end algorithm;

thread implementation algorithm.cc
end algorithm.cc;

data logs
end logs;

data implementation logs.error_logs
end logs.error_logs;

process monitor
features
error_data_in: requires data access logs.error_logs
                    { Access_Right => read_only;};
end monitor;

process implementation monitor.error_monitor
subcomponents
comm_errors: thread m_algorithm.errors;
end monitor.error_monitor;

thread m_algorithm
features
c_error_data: requires data access logs.error_logs
  { Access_Right => read_only;};
end m_algorithm;

thread implementation m_algorithm.errors
end m_algorithm.errors;
```

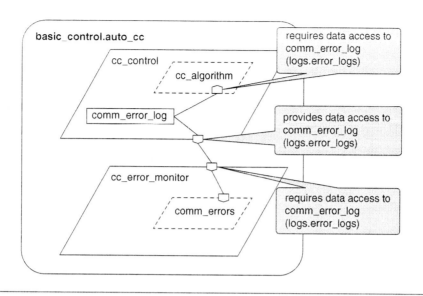

10.3 Bus Access and Connections

The computer hardware components and physical components of your embedded system are physically connected to each other via buses. You accomplish this by declaring the need for a component such as a processor or sensor device to access a bus of a certain type, and making the access connection through bus access connection declarations. For example, if a processor requires access to a PCI bus, the connection of the processor's requires bus access feature to the bus establishes a physical connection between them. Similarly, a device such as a digital camera may have a requires bus access feature of bus type *usb* that is connected to a bus of type *usb*. Bus access connections can be directional or bidirectional. As in the case of the data access connection, the direction indicates transfer to or transfer from the bus and must be compatible with the *Access_Right* property values of the source and destination.

The syntax for the bus access features and bus access connections is the same as for data access with the keyword *data* replaced by *bus*. Listing 10-8 shows an example of bus access connections for a simplified cruise control system that consists of a cruise control unit (system component) and pilot input, speed sensor, and throttle devices.

Additional execution hardware for the system consists of a processor that executes the application software and a bus connecting the hardware components. A complete path for a bus access connection is between a bus and a terminating requires bus access feature (i.e., the bus access feature of the component to be connected to the bus). This may progress through multiple access features both provides and requires. The figure in the lower portion of Listing 10-8 is a graphical representation for required access features and connections to the bus declared in the text. Several devices and a processor are connected to a *CANBus* within the cruise control system. The *CANBus* is also made accessible to other components outside the cruise control system. Only the connection declaration from the *CANBus* to the *external_access* feature for the system *cruise_control_system.impl* is included in the example.

Listing 10-8: *Basic Bus Access and Access Connection Declarations*

```
system cruise_control_system
features
external_access: provides bus access CANBus.impl;
end cruise_control_system;

system implementation cruise_control_system.impl
subcomponents
pilot_input_unit: device pilot_input_unit;
speed_sensor: device speed_sensor;
CCU: system CCU_system;
throttle_actuator: device  throttle_actuator;
M555: processor M555;
CANBus: bus CANBus.impl;
connections
-- bus access connections
bus_access_01: bus access  CANBus <->  pilot_input_unit.bus_access;
bus_access_02: bus access  CANBus <->  speed_sensor.bus_access;
bus_access_03: bus access  CANBus <->  throttle_actuator.bus_access;
bus_access_04: bus access  CANBus <-> M555.bus_access;
bus_access_05: bus access  CANBus <-> external_access;
end cruise_control_system.impl;

device pilot_input_unit
features
bus_access: requires bus access CANBus.impl;
end pilot_input_unit;

bus CANBus
end CANBus;
```

```
bus implementation CANBus.impl
end CANBus.impl;

device speed_sensor
features
bus_access: requires bus access CANBus.impl;
end speed_sensor;

device throttle_actuator
features
bus_access: requires bus access CANBus.impl;
end throttle_actuator;

processor M555
features
bus_access: requires bus access CANBus.impl;
end M555;
```

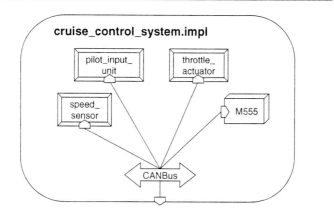

Listing 10-9 illustrates how to model two subsystems with hardware components and bus connections, where their local buses are interconnected via a backbone. Some of the specifications are not complete (e.g., type rather than implementation classifiers are used in defining some of the components and subcomponents). In the illustration, each subsystem uses a local 1553 bus to connect its internal devices. The two subsystems are interconnected via an Ethernet backbone. We do this by connecting each of the local buses directly to the backbone via *requires bus access* features on the local buses. The bus access, requires, provides, and connections are shown both graphically (lower portion of Listing 10-9) and as AADL text declarations.

Listing 10-9: *Example Bus Access Connection Declarations*

```
system vehicle_system
end vehicle_system;

system implementation vehicle_system.impl
subcomponents
bow: system frontend.impl;
stern: system backend.impl;
backbone: bus Ethernet;
connections
bus access backbone <-> bow.network_bus;
bus access backbone <-> stern.network_bus;
end vehicle_system.impl;

system frontend
features
network_bus: requires bus access Ethernet;
end frontend;

system implementation frontend.impl
subcomponents
B_1553: bus B_1553;
PC1: processor PC;
Obstacle_Sensor: device IR_Sensor;
connections
C01: bus access B_1553.network_bus <-> network_bus;
C02: bus access B_1553 <-> PC1.LocalBus;
C03: bus access B_1553 <-> Obstacle_Sensor.LocalBus;
end   frontend.impl;

system backend
features
network_bus: requires bus access Ethernet;
end backend;

system implementation backend.impl
subcomponents
PC2: processor PC;
B_1553: bus B_1553;
Depth_Sensor: device UltraSound_Sensor;
connections
C01: bus access B_1553.network_bus <-> network_bus;
C02: bus access B_1553 <-> PC2.LocalBus;
C03: bus access B_1553 <-> Depth_Sensor.LocalBus;
end backend.impl;

processor PC
features
LocalBus: requires bus access B_1553;
end PC;
```

```
bus B_1553
features
network_bus: requires bus access Ethernet;
end b_1553;

device IR_Sensor
features
LocalBus: requires bus access B_1553;
end IR_Sensor;

device UltraSound_Sensor
features
LocalBus: requires bus access B_1553;
end UltraSound_Sensor;

bus Ethernet
end Ethernet;
```

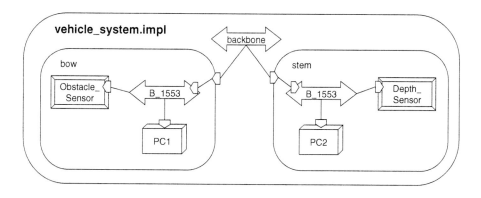

10.4 Feature Groups and Connections

Feature groups represent collections of component features or other feature groups. You can specify the makeup of a feature group by a feature group type. You can connect feature groups using a single connection declaration. From within a component, features contained within a feature group can be connected to individually. From outside a component, feature groups can be connected as a single unit, where this single connection represents multiple feature connections. Feature groups allow the number of connection declarations to be reduced, especially at higher levels of a system, when multiple features from one subcomponent and its contained subcomponents must be connected to features in another subcomponent and its contained subcomponents.

You can group features with common properties (e.g., all event ports, all bus accesses) or mix feature types and directions within a feature group. When a feature group is declared for a component, it may be only partially specified by leaving off the feature group type, or by referring to a feature group type for which the features have not been filled in yet. This is useful when you want to model interaction between subsystems early in the development without knowing the exact details of the interactions.

10.4.1 Declaring Feature Group Types

The content of a feature group is defined in a feature group type declaration that explicitly identifies the individual features and feature groups that comprise it. There is no implementation declaration for a feature group. Feature group types can be declared for any kind of feature and may consist of one type of feature (e.g., ports) or a composite of a variety of feature types. Feature group type declarations are placed in packages separate from or together with component type and implementation declarations.

The declaration is similar to a component type declaration with a **features** and a **properties** section. As with other component type declarations, properties of the feature group can be declared and a feature group type can be extended and refined.

Three examples of **feature group** type declarations are shown in Listing 10-10. The feature group *roll_set* includes a mix of features including data and event ports, a bus access, and the feature group *error_set*. The feature group *error_set* includes a requires access declaration for the data component *error_data*. Two input ports are included as the features for the feature group *dual_Inports*.

Listing 10-10: *Example Feature Group Type Declarations*

```
feature group roll_set
features
roll_data: in data port;
roll_cmd: out data port;
engage: in event port;
h_bus: requires bus access;
errors: feature group error_set;
end roll_set;

feature group error_set
features
sensor_error: in data port;
range_error: out event data port;
```

```
error_log: requires data access error_data;
end error_set;

feature group dual_Inports
features
input1: in data port;
input2: in data port;
end dual_Inports;
```

A feature group type can be declared as the inverse of another feature group type. This is useful for matching the type of the source and destination of feature group connections, as we will see in Section 10.4.3. This relationship is indicated by the reserved words **inverse of** and the name of a feature group type. You may declare a feature group type as inverse without explicitly listing the features. In this case, the features are known by the names declared in the opposite feature group type. The features of the inverted feature group must be in the same order as in the referenced feature group but with the opposite directions. A feature group type that is named in an **inverse of** statement cannot itself contain an **inverse of** statement. Thus, a chaining of inverses, such as B **inverse of** A and C **inverse of** B, is not permitted.

Feature group type declarations using **inverse of** are shown in Listing 10-11. The feature group *control_plug* is simply the inverse of the feature group *control_socket*. The feature group type *control_plug_ independent* has declared the features with different names; they are matched in order with the features of the inverse feature group type *control_socket*.

Listing 10-11: *A Feature Group Type Declaration and Its Inverse*

```
feature group control_socket
features
  Wakeup: in event port;
  Observation: out data port position;
end control_socket;

feature group control_plug
inverse of control_socket
end control_plug;

feature group control_plug_independent
features
Activate: out event port;
Measurement: in data port position;
inverse of control_socket
end control_plug_independent;
```

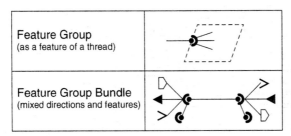

Feature Group (as a feature of a thread)	
Feature Group Bundle (mixed directions and features)	

Figure 10-8: *Graphical Representations of Feature Groups*

Figure 10-8 contains graphical icons for feature groups and their connections. Feature groups can bundle different features and port types and directions as shown.

10.4.2 Declaring a Feature Group as a Feature of a Component

To declare a feature group as a feature of a component type, you use the declaration template shown in the following box.

```
name : feature group [ [ inverse of ] feature group type]
  [{properties}] ;
```

The optional feature group type reference defines the contents of the feature group for the component. Initially, this reference may be omitted in the declarative model. The feature group may refer to a feature group type that initially does not have any features. This allows you to model interactions between two subsystems without knowing the details of the interaction through a collection of ports and other features. You can declare the feature group to be the inverse of a feature group type. This indicates that it represents the features in the feature group type, but with opposite direction. This is useful when matching the feature group types of feature group connections (see Section 10.4.3). By using **inverse of** in the feature declaration you can avoid having to declare two feature group types, one for each end of a port group connection.

Examples of feature group declarations are shown in Listing 10-12. These examples reference the feature group types shown in Listing 10-10 and are excerpts from a complete specification consisting of the relevant declarations and portions of declarations needed to show what is required in specifying a specific feature group. Note that feature group declarations can be mixed with other feature declarations as shown for the thread *read_data*.

Listing 10-12: *Sample Feature Group Declarations*

```
system control
features
roll_01: feature group roll_set;
end control;

thread read_data
features
inputs: feature group inverse of dual_Inports;
output: out data port;
end read_data;

thread monitor
features
error_set: feature group error_set;
end monitor;
```

10.4.3 Declaring Feature Group Connections

Connections can be made between feature groups of two subcomponents, or between the individual features within a feature group of an enclosing component and features of subcomponents. The declaration template for a feature group connection is shown in the following box.

```
name : feature group <source feature group> <->
  <destination feature group> [{properties}] [in modes];
```

Elements of a feature group of a component can be individually connected to features of subcomponents within that component. However, elements of a feature group of a subcomponent cannot be referenced outside the subcomponent (i.e., they cannot be connected to feature group elements or ports of other subcomponents). In other words, grouping and pulling apart elements of a feature group can occur when going up or down the component hierarchy, but not within the same level of the component hierarchy. This is shown topologically in Figure 10-9, where the feature group *gc_data* has individual connections to ports on the subcomponents of the process *GandC*. In addition there is a single connection from the feature group *c_data* to the feature group *c_data_inv* between two threads of the process *GandC*.

Excerpts from an AADL model for a simple cruise control system are shown in Listing 10-13. They include a sample feature group connection for the feature group *mode_control_group* and its inverse. The feature group *mode_control_group* combines all of the event ports emanating from a data processing component that impact the modal

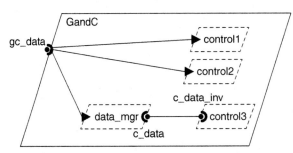

Figure 10-9: *Individual Feature Group Connections*

behavior of a control system. The single feature group connection *d_to_c* is shown within the system implementation *subsystem.cc_process_ subsystem* between the process *process_raw_data* and the process *controller*. The individual connections from each of the event ports in the feature group *mc_in* (an instance of the inverse of the feature group *mode_control_group*) to in event ports on the three subcomponent threads of the process *controller*. A graphical representation of these connections is shown in the lower portion of Listing 10-13. Individual ports and connections are not labeled in the graphic.

Listing 10-13: *Sample Feature Group Connection Declarations*

```
system subsystem
end subsystem;

system implementation subsystem.cc_process_subsystem
subcomponents
process_raw_data: process process_data.cc_process_raw_data;
controller: process control.cc_control;
connections
d_to_c: feature group process_raw_data.mc_out <->
                    controller.mc_in;
end subsystem.cc_process_subsystem;

process process_data
features
mc_out: feature group mode_control_group;
end process_data;

process control
features
mc_in: feature group mode_control_group_inverse;
end control;

thread controller
features
```

```
mode_event: in event port;
end controller;

process implementation control.cc_control
subcomponents
control1: thread controller;
control2: thread controller;
control3: thread controller;
connections
C1: port mc_in.cc_on_out01 -> control1.mode_event;
C2: port mc_in.cc_off_out01 -> control2.mode_event;
C3: port mc_in.brake_on_out01 -> control3.mode_event;
end   control.cc_control;

process implementation process_data.cc_process_raw_data
end process_data.cc_process_raw_data;

feature group mode_control_group
features
cc_on_out01: out event port;
cc_off_out01: out event port;
brake_on_out01: out event port;
end mode_control_group;

feature group mode_control_group_inverse
inverse of mode_control_group
end mode_control_group_inverse;
```

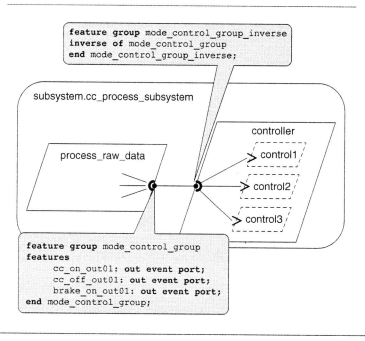

The feature group types of connected feature groups must match. By default, this means that the feature group types must be identical for connections from a subcomponent to the enclosing component, and they must be inverses of each other on connections between subcomponents (i.e., they must have the same set of features but with opposite direction).

It is possible to have two other matching rules: *equivalence/complement* and *subset*. *Equivalence/complement* is used when feature group types have been defined independently by different teams and cannot be modified by the integrator to add an inverse of indicator. *Subset* is used to allow a destination feature group to be a subset of the source feature group, as is the case when modeling the OMG Data Distribution Service [DDS].

The connection matches are specified by the *Classifier_Matching_Rule* property. The standard values of this property for feature group connections up or down the component hierarchy and for feature group connections between subcomponents are summarized in Table 10-4.

Feature groups can be effective in organizing data and connections. For example, you can bundle the individual outputs of multiple sensors within a sensor subsystem network into a single feature group. In that instance, all of the sensor data is transferred through a single connection declaration from the sensor subgroup to a control processing system.

Table 10-4: *Classifier Matching Rules for Feature Group Connections*

Values for Connections Through the Component Hierarchy	
Value	**Description**
Classifier_Match	Source type is identical to the destination type (the default value).
Equivalence	Classifiers of a connection are considered to match when listed in the Supported_Classifier_Equivalence_Matches property as pairs of classifier values that define acceptable matches. Either element of the pair can be the source or destination classifier. (Equivalence is used when the two types are identical such that their elements match.).
Subset	Classifiers of a connection are considered matching if the outer feature group has outgoing features that are a subset of outgoing features of the inner feature group, and if the inner feature group has incoming features that are a subset of incoming features of the outer feature group. The pairs of features must have the same name.

Table 10-4: *Classifier Matching Rules for Feature Group Connections (continued)*

Values for Connections Between Subcomponents	
Value	**Description**
Classifier_Match	Source type is the complement of the destination type (the default value).
Complement	Classifiers of a connection are considered to be complementary when listed in the Supported_Classifier_Complement_Matches as pairs of classifier values that define acceptable matches. Either element of the pair can be the source or destination classifier. (Complement is used when the two types are identical such that their elements match.)
Subset	Classifiers of a connection are considered to match if each has incoming features that are a subset of the outgoing features of the other. The pairs of features must have the same name.

When you instantiate AADL models with feature group connections a separate connection instance is created for each feature in the feature group. If you are instantiating a partial model where feature groups have no feature group type or an incomplete feature group type, then a connection instance is created between the feature groups. In other words, you can analyze partial models early in the development process (e.g., you can determine the workload generated on a network by interactions between subsystems on different processors based on a property of the feature group connection indicating an anticipated bandwidth or bandwidth budget). More information on connections in instance models can be found in Section 4.1.6.

10.5 Abstract Features and Connections

It is possible for you to define abstract (generic) features and connections among them and later replace these abstractions with concrete features and connections (such as ports and port connections). You will find this capability useful for defining component templates that you can partially define and apply in a specific model. For example, you can use a feature on thread component type and refine it to a data port for a specific thread instance. Similarly, you can define an abstract component with multiple abstract features and refine it into a process component with the abstract features refined into the ports and access features needed for a model. This capability can also be used in developing a conceptual model such that an interface is defined but the specific data or transfer mechanisms through that interface are not known.

10.5.1 Declaring Abstract Features

You declare abstract features in component types using the format shown in the following box.

```
name: [direction]  feature [feature prototype];
```

The optional direction can be either **in** or **out** but not **in out**. If the direction is not specified the abstract feature can be refined into a directional or bidirectional feature. If the direction is specified for an abstract feature, then the refinement of this feature must be compatible with the specified feature. In the case of access features, the direction reflects the read or write.

Listing 10-14 shows abstract features within the system type *dynamic_control_partial*. The abstract feature *sensor_data* is declared as an **in feature** with a data type classifier *sensor_format*. This can be used in the situation where the data content is known but the decision has not been made as to whether the data will be transmitted via data port or event data port (message) communication. The feature *error_report* is defined as an **out feature** without a constraining data classifier to report errors as events (using event ports) or as messages (using event data ports). In contrast, the out abstract feature *command* is defined with a feature prototype identifier *cmd* that is declared in the prototypes section of *dynamic_control_partial*.

Listing 10-14: *Abstract Feature Declarations*

```
system dynamic_control_partial
prototypes
  cmd: out feature;
features
    sensor_data: in feature sensor_format;
    error_report: out feature;
    command: out feature cmd;
end dynamic_control_partial;
```

10.5.2 Refining Abstract Features

You refine an abstract feature in one of two ways: by a feature refinement declaration or by supplying an appropriate prototype actual. Listing 10-15 shows a declaration of *dynamic_control* that extends the system type *dynamic_control_partial*. In this extension, the abstract feature *sensor_data* is declared to be an in data port using **refined to**, reflecting the decision to use synchronized communication via data ports rather than asynchronous message passing. Similarly, the abstract feature *error_report* is **refined to** an out event port, reflecting the decision to have only error event reporting without data.

The **out feature** *command* is refined via a prototype binding of the feature prototype *cmd* to be an **out event data port** with the data type *cmd_format*. It is not necessary to repeat the direction in the prototype binding or to use an explicit **refined to** declaration for the abstract feature *command*. This refinement captures the decisions to use message passing and a defined data type *cmd_format* for command output. The data type *cmd_format* is shown in the listing.

Listing 10-15: *Abstract Feature Refinements*

```
system dynamic_control extends dynamic_control_partial
                    (cmd => out event data port cmd_format)
features
    sensor_data: refined to in data port sensor_format;
    error_report: refined to out event port;
end dynamic_control;

data cmd_format
end cmd_format;

data sensor_format
end sensor_format;
```

10.6 Arrays and Connections

When you declare connections, the end-points of the connections may be subcomponent arrays. In this case, the connection represents a collection of connections between the different elements of the arrays. You can specify the desired set of connections in two ways: you can provide a list of index pairs into both arrays to represent each connection; or you can specify a connection pattern.

You can declare a one-dimensional array of threads inside a process implementation that is used in a one-dimensional process array declaration. Similarly, you can declare a one-dimensional array of devices in a system implementation to represent a sensor array on a board, for which you have multiple instances declared as an array. In both of these cases, you effectively need to connect a two-dimensional set of threads and a two-dimensional set of devices.

You can also connect a feature of a component array with a feature array of a single component. In this case, the resulting set of connections connects the feature of different component array elements to different feature array elements of the other component. Feature arrays can only be declared as one-dimensional arrays. This connection pattern is useful when you need to model data concentrators, voters in redundant systems, and health monitors for large-scale systems such as sensor nets.

10.6.1 Explicitly Specified Array Connections

Connections for subcomponent arrays may be declared explicitly as shown in Listing 10-16. You can do this by specifying a *Connection_Set* property that lists the indices of the source and destination arrays as pairs of values, one for each desired connection. In the example, the out data port on the first device subcomponent of the *sensor_network* array is connected to the first two of the elements of the *data_filters* process array, and the second device is connected to the second two elements of the *data_filters* process array.

Listing 10-16: *Explicit Array Connections*

```
subcomponents
sensor_network: device radar_sensor [2];
data_filters: process filters.basic [4];
connections
conn_1: port sensor_network.output01 -> data_filters.input_data
  { Connection_Set => ( ( 1, 1), ( 1, 2), (2, 3), (2, 4)); };
----------
device radar_sensor
features
output_01: out data port;
output_02: out data port;
end radar_sensor;

process filters
features
input_data: in data port;
end filters;

process implementation filters.basic
end filters.basic
```

10.6.2 Array Connection Patterns

The *Connection_Pattern* property specifies how port connections are replicated if the ultimate source and final destination for a connection are arrays. This allows you to express common connection patterns quickly without listing them out explicitly. The standard values for the *Connection_Pattern* property are shown in Table 10-5. These can be combined by listing multiple pattern values for the *Connection_Pattern* property as shown in Figure 10-11. If your system has connection patterns that are not expressible by the pattern property, then you can introduce your own pattern extension and generate the explicitly declared connections as shown in Section 10.6.1.

Table 10-5: *Standard Values of the Connection_Pattern Property*

Value	Description
All-To-All	Each array element of the source has a semantic connection to each element in the destination.
One-To-One	Elements of the source array and the destination array have pair wise semantic connections. This property value applies if the two arrays have identical ranges or if one range is a subset of the other then only the subset starting with the first element is connected.
Next or Previous	Elements of the source array are connected to the next (previous) element in the destination array. The last element does not connect in the case of next and the first element does not connect in the case of previous. This property value applies if the two arrays have identical ranges.
Cyclic_Next or Cyclic_Previous	Elements of the source array are connected to the next (previous) element in the destination array. In the case of *Cyclic_Next* the last element in the array connects to the first, and vice versa for *Cyclic_Previous*. This property value applies if the two arrays have identical ranges.
One-to-All	A single element of the source has a semantic connection to each element in the destination. This connection pattern is used when the destination array has a higher dimensionality than the source array.
All-to-One	Each array element of the source has a semantic connection to a single element in the destination. This connection pattern is used when the destination array has a lower dimensionality than the source array.

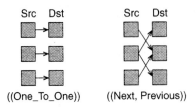

Figure 10-10: *Connection Patterns for Single Dimension Arrays*

Figure 10-10 shows two example topologies for connections between two single-dimensional arrays consisting of three elements. In these examples, the patterns of the connections are shown without explicitly including features on the components. In the first pattern, each *Src* element is connected to the *Dst* element with the same index. In the second pattern, an *Src* element of a specific index is connected to the *Dst* elements with one index higher and one index lower.

You can also define connections within the same component array. Figure 10-11 illustrates the use of connection patterns in a two-dimensional array for some simple patterns. In the first pattern the first (down) dimension specifies *One_To_One* and the second dimension *Next*. In the second pattern both dimensions follow the *cyclic_Next* pattern.

10.6.3 Using Array Connection Properties

Consider a system where a network of identical temperature sensors provides data to an array of processes. The process array delivers commands to temperature control actuators. Each temperature sensor has an output port, each process has an in port and an out port, and each temperature control actuator has an in port. Rather than declaring individual connections among these components, you can use the *Connection_Pattern* to specify the connections. The topology and declarations for this system are shown in Listing 10-17, where each sensor is connected to each of the processes ((all-to-all)) and each process provides a command to one actuator ((one-to-one)).

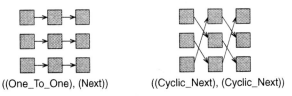

Figure 10-11: *Connection Patterns Within a Two-Dimensional Component Array*

Listing 10-17: *Example of Connecting Subcomponent Arrays*

```
device temp_sensor
features
output: out data port;
end temp_sensor;

device temp_control_actuator
features
input: in data port;
end temp_control_actuator;

process control
features
input: in data port;
output: out data port;
end control;

system control_system
end control_system;

system implementation control_system.temperature
subcomponents
sensor_array: device temp_sensor [5];
actuator_array: device temp_control_actuator [5];
process_array: process control [5];
connections
dpc1: port sensor_array.output -> process_array.input
    { Connection_Pattern=> ((All_To_All)) ;} ;
dpc2: port process_array.output -> actuator_array.input
    { Connection_Pattern=> ((One_To_One)) ;};
end control_system.temperature ;
```

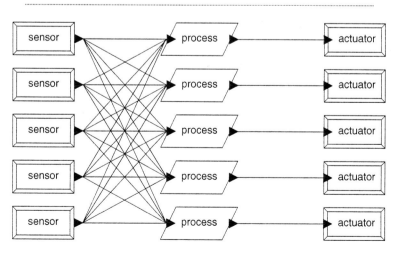

10.7 Subprogram Calls, Access, and Instances

With AADL, you can model the execution behavior within a thread in several ways. In this section, we show you how you can use subprogram call sequences, subprogram access, and subprogram subcomponents to model both local and remote calls. You also have the opportunity to describe the thread behavior through the Behavior Annex standard of AADL [BAnnex].

You declare calls to subprograms via named sequences in the **calls** section of a thread or subprogram implementation. In these sequences, each entry in a call sequence is a call declaration. A call sequence can contain one or more call declarations. The template for a call sequence is shown in the following box. You can call subprograms in subprogram groups by identifying a **provides subprogram access** feature in a subprogram group classifier. You can also call subprogram features in data components by identifying the data component type and the **provides subprogram access** feature.

```
calls
Call_Sequence_Name: {
            CallName : subprogram [subprogram reference]
[{properties}] [in modes];
            call_declaration2
    };
```

You can model subprogram calls in three ways: by identifying the subprogram classifier of the subprogram to be called; by modeling remote subprogram calls as a binding to the remote subprogram or as a subprogram access connection; and by explicitly modeling subprogram instances and identifying a specific instance in the call.

If classifier references are used, the specific instance that is called is defined through a subprogram access connection that establishes a path from the subprogram instance to the calling component. Alternatively, a subprogram call may name a subprogram classifier. In this case, the call is to an implicit local instance of the subprogram. You will use this option if you just want to identify the subprogram.

If a thread declares a **provides subprogram access** feature, it indicates that another thread can call this subprogram remotely. AADL supports two ways of specifying remote procedure calls. The first way to specify such a remote call is by attaching an *Actual_Subprogram_Call* property to the call that simply refers to the remote thread and subprogram feature. This corresponds to lining in a proxy library for the remote call and identifying the server to be called. The second way of specifying the remote call is to declare subprogram access connections from the **requires subprogram access** feature of the calling thread to the provides subprogram access feature of the thread being called. This follows the component-based development paradigm where all interactions with other components are reflected in the component interface, including calls to subprograms.

If you want to explicitly model instances of subprograms, you can declare subprogram subcomponents, and name them in the call. If the subprogram subcomponent exists outside a thread (e.g., is shared by all threads of a process) then the thread type can be specified with a requires subprogram access feature and this access feature is named by the call. You will model subprogram instances if you want to explicitly record the fact that your application build contains multiple copies of the same source code or source code library instead of expecting the linker/loader tool to provide such information back to the model.

10.7.1 Declaring Calls and Call Sequences

An example call sequence is shown in the partial specification of Listing 10-18 where the call sequence *two_calls* involves a call to the subprogram implementations *acquire.temp* and then *adjust.level*. The associated subprogram declarations are also shown. The call sequence is determined by the subprogram calls declaration order. In other words, the calls order is linear. If more complex call orderings are desired, an annex notation could provide a specification of other orderings, such as a "branch" or "iteration." Alternatively, one can specify different calls sequences that are active under different modes. Notice that subprograms may call other subprograms. This is shown in Listing 10-18 where the subprogram implementation *adjust.level* calls the subprogram *find.temp_values*. Graphically, subprogram calls are represented by subprogram symbols, arranged left to right within a thread implementation or subprogram symbol. A call sequence arrow may be included as shown in the figure in the lower portion of Listing 10-18.

Listing 10-18: *Example Subprogram Calls*

```
thread implementation control.thermal_control
--
calls
two_calls:{
                get_temp: subprogram acquire.temp;
                adjust_level: subprogram adjust.level;
        };
```

continues

```
--
end control.thermal_control;
subprogram acquire
end acquire;
subprogram implementation acquire.temp
end acquire.temp;
subprogram adjust
end adjust;
subprogram implementation adjust.level
calls
call_seq: {
          find_scale_values: subprogram find.temp_values;
        };
end adjust.level;
subprogram find
end find;
subprogram implementation find.temp_values
end find.temp_values;
```

10.7.2 Declaring Remote Subprogram Calls as Bindings

You can use properties to specify remote subprogram calls, as shown for client-server interactions in Listing 10-19. The property association *Actual_Subprogram_Call* declares that the subprogram call *call_server* within the thread *calling_thread*, which is a subcomponent of the process *client_process*, is being made to the subprogram contained within the server process (*server_process*).

Listing 10-19: *Client-Server Subprogram Call Binding Example*

```
system implementation client_server_sys.impl
subcomponents
client_process: process client_process.impl;
server_process: process server_process.impl;
```

```
properties
Actual_Subprogram_Call =>
    reference(server_process.server_thread.service)
    applies to  client_process.calling_thread.call_server;
end client_server_sys.impl;
--
process client_process
end client_process;
--
process implementation client_process.impl
subcomponents
calling_thread: thread calling.impl;
end client_process.impl;
--
thread calling
end calling;
--
thread implementation calling.impl
calls   server_call_sequence:
        {
            call_server: subprogram service_it ;
        };
end calling.impl;
----
process server_process
features
service: provides subprogram access service_it;
end server_process;
--
process implementation server_process.impl
subcomponents
server_thread: thread server_thread.impl;
end server_process.impl;
--
thread server_thread
features
service: provides subprogram access service_it;
end server_thread;
--
thread implementation server_thread.impl
end server_thread.impl;
--
subprogram service_it
end service_it;
```

continues

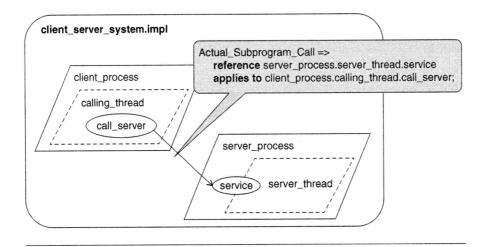

10.7.3 Declaring Remote Subprogram Calls as Access Connections

You can explicitly declare a path from the subprogram call to the remote subprogram in a thread, as shown in Listing 10-20. You add requires subprogram access features to the client side and provides subprogram access features to the server side. You then declare subprogram access connections to represent the call connection sequence.

Listing 10-20: *Client-Server Subprogram Access Connection Example*

```
system implementation client_server_sys.impl
subcomponents
client_process: process client_process.impl;
server_process: process server_process.impl;
connections
subprogram access client_Process.service <->
    server_process.service;
end client_server_sys.impl;
--
process client_process
features
service: requires subprogram access service_it;
end client_process;
--
process implementation client_process.impl
subcomponents
calling_thread: thread calling.impl;
connections
subprogram access calling_thread.service <-> service;
end client_process.impl;
```

```
--
thread calling
features
service: requires subprogram access service_it;
end calling;
--
thread implementation calling.impl
calls   call_seq: {
            call_server: subprogram service ;
        };
end calling.impl;
----
process server_process
features
service: provides subprogram access service_it;
end server_process;
--
process implementation server_process.impl
subcomponents
server_thread: thread server_thread.impl;
connections
service_conn: subprogram access server_thread.service <->
    service;
end server_process.impl;
--
thread server_thread
features
service: provides subprogram access service_it;
end server_thread;
--
thread implementation server_thread.impl
end server_thread.impl;
--
subprogram service_it
end service_it;
```

10.7.4 Modeling Subprogram Instances

You model subprogram instances by declaring them as subprogram subcomponents or as subprogram group subcomponents. This allows you to have an explicit record in the AADL model of the number of instances of source code and source code libraries. You can record the results from the linker/loader as source code size properties on processes to reflect the size of the load image.

You can combine explicit modeling of subprogram instances with explicitly recording the call to a specific subprogram instance by naming it in the call, instead of naming the classifier. In this case, you make use of subprogram access features as illustrated in Section 10.7.3, if the subprogram instance is located in another component.

Examples of these declarations are shown in Listing 10-21. There is a local access connection from the subprogram *verify_value* to the *v_acc* requires subprogram access feature. This defines the subprogram that is called by the call *verify_local* in the thread *crl_laws*. The remote subprogram access connection that extends from the subprogram *scale_acc_data* to the feature *r_s_d_acc* establishes the subprogram *scale_acc_data* as the subprogram instance that is called in the *scale_remote* call of the thread *crl_laws*. The properties *Input_Rate* and *Output_Rate* can be used to specify the call rates for subprogram calls.

Listing 10-21: *Example Subprogram Access Connections*

```
subprogram scale_data
end scale_data;

subprogram implementation scale_data.sensor
end scale_data.sensor;

subprogram verify
end verify;

subprogram implementation verify.basic
end verify.basic;

thread ctrl
features
r_s_d_acc: requires subprogram access scale_data.sensor;
v_acc: requires subprogram access verify.basic;
end ctrl;

thread implementation ctrl.basic
calls
basic_calls:{
            verify_local: subprogram v_acc;
            scale_remote: subprogram r_s_d_acc;
        };
end ctrl.basic;
--

data accelerometer
features
s_d_acc: provides subprogram access scale_data.sensor;
end accelerometer;
```

```
data implementation accelerometer.basic
subcomponents
scale_acc_data: subprogram scale_data.sensor;
connections
sac1: subprogram access scale_acc_data <-> s_d_acc;
end accelerometer.basic;

process sensor
features
s_d_acc: provides subprogram access scale_data.sensor;
end sensor;

process implementation sensor.basic
subcomponents
accelerometer_basic: data accelerometer.basic;
connections
sa1: subprogram access accelerometer_basic.s_d_acc <->
    r_s_d_acc;
end sensor.basic;

process ctrl_process
features
r_s_d_acc: requires subprogram access scale_data.sensor;
end ctrl_process;

process implementation ctrl_process.basic
subcomponents
ctrl_laws: thread ctrl.basic;
verify_value: subprogram verify.basic;
connections
ac1: subprogram access verify_value <-> ctrl_laws.v_acc;
ac2: subprogram access r_s_d_acc <-> ctrl_laws.r_s_d_acc;
end ctrl_process.basic;

system basic
end basic;

system implementation basic.control
subcomponents
ctrl_process_basic: process ctrl_process.basic;
sensor_basic: process sensor.basic;
connections
c1: subprogram access sensor_basic.s_d_acc <->
ctrl_process_basic.r_s_d_acc;
end basic.control;
```

continues

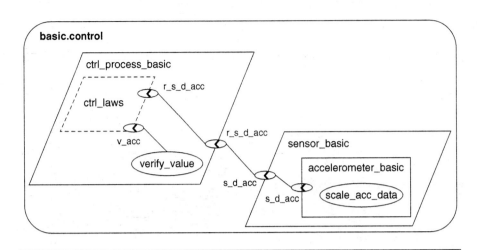

10.8 Parameter Connections

Parameters represent data values passed into and out of a subprogram. Parameter connections are used to describe the flow of data from a thread into and out of a subprogram and the data flow between called subprograms in a subprogram call sequence within a thread or within subprograms. You are not required to model this level of detail in an architecture model, but AADL gives you the capability to represent the flow of data between subprogram calls and between a subprogram call and the incoming or outgoing port of the thread making the call.

10.8.1 Declaring Parameters

Parameters are declared as features in a type declaration of a subprogram. The declaration is similar to data port declarations, except the reserved word **parameter** rather than **data port** is used. The template for a parameter declaration is shown in the following.

```
name: <direction>  parameter [data identifier] {properties} ;
```

The direction parameter entry can be in, out, or in out. The optional data identifier that can be included for parameters defines the nature of the data that is passed into or out of the subprogram. This may be the name of a data type or a data implementation or may be omitted.

Example parameter declarations are shown in Listing 10-22. Note that data type and implementation identifiers are included for some of the declarations (e.g., *raw_data, raw_data.interim*). The property association for *Source_Name* defines the name of the parameter in the source for the subprogram.

The parameter feature represents call-by-value parameters (i.e., the data value is passed into and out of a subprogram). If you need to represent call-by-reference parameter, you represent it as a *requires data access* feature of the subprogram.

Listing 10-22: *Example Parameter Declarations*

```
subprogram scale
features
in_parameter: in parameter raw_data;
interim_value: out parameter raw_data.interim;
end scale;
--
subprogram edit_range
features
interim_value: in parameter;
out_parameter: out parameter;
end edit_range;
--
subprogram update_set
features
io_parameter: in out parameter { Source_Name => "IOdata";};
end update_set;
```

There are two standard properties that apply to parameters. They specify aspects of the source representations associated with the parameters. The *Source_Name* property specifies the name within the associated source code that corresponds to the parameter. The *Source_Text* property specifies a source code file or list of files that contain the parameter. Other properties such as the *Source_Data_Size* are derived from the properties of the data classifier referenced by the parameter declaration.

10.8.2 Declaring Parameter Connections

Listing 10-23 presents textual and graphical representations of the parameters and the parameter connections associated with a calls sequence within a thread. Parameter connections are always directional. In a graphical representation

- Parameters are represented as solid arrows (▶), like data ports.
- Parameter connections are shown as solid lines (—) between parameters or between a parameter and a port (on a containing thread of the subprogram call).
- Subprogram calls are represented by ovals (◯) labeled with the call (e.g., *scale*) and called subprogram type.
- A call sequence is indicated by an arrow with an open arrow head (→). (Alternatively, a call sequence can be specified by the ordering of the calls from the left to the right.)

Notice that the in event data port *in_data* of the thread *scale_data* is connected to the parameter *in_parameter* of the subprogram scale. Parameters can be connected to data ports and event data ports.

Listing 10-23: *Example Parameter Connections*

```
thread scale_data
features
in_data: in event data port;
out_data: out data port;
end scale_data;
--
thread implementation scale_data.impl
calls    call_seq: {
scale: subprogram scale;
edit: subprogram edit_range;
update: subprogram update_set;
        };
connections
pconn1: parameter in_data -> scale.in_parameter;
pconn2: parameter scale.interim_value -> edit.interim_value;
pconn3: parameter edit.out_parameter -> update.io_parameter;
pconn4: parameter update.io_parameter -> out_data;
end scale_data.impl;
--
subprogram scale
features
in_parameter: in parameter;
interim_value: out parameter;
end scale;
--
subprogram edit_range
features
interim_value: in parameter;
out_parameter: out parameter;
end edit_range;
```

```
--
subprogram update_set
features
io_parameter: in out parameter;
end update_set;
```

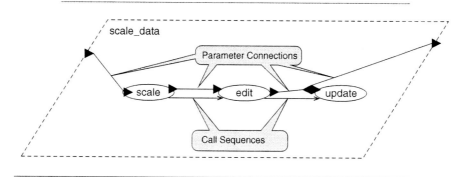

Chapter 11

System Flows and Software Deployment

You have learned how to describe your system in terms of a hierarchy of components, and define the interaction between components in terms of connections. In this chapter, you will learn how to specify system level information. First, you will learn how to specify end-to-end flows through the system. This allows you to define critical information flows. They can reflect use scenarios and be the basis for system-level analyses. Second, you will learn how to bind the embedded application software onto the computer platform. This binding represents a deployment configuration, which is essential for any analysis that involves operational quality attributes, such as performance, security, or reliability.

11.1 Flows

Flows enable the representation and analysis of logical paths through an architecture. In support of flows, you can declare flow specifications for components. A flow specification defines a logical flow from a component input to a component output. These flows can be through ports as well as data access features and can represent any logical flow, such as data flow, control flow, or fault event flow. You can attach flow-related properties, such as latency or accuracy that can be used in end-to-end flow analysis without having to examine the implementation of the component. In component implementations you elaborate those flow specifications into flow implementations (i.e., as paths through its subcomponents). These paths start with the incoming feature of the flow specification, follow a sequence of connections and subcomponent flow specifications, and end with the outgoing feature of the flow specification. Similarly, you can define end-to-end flows. They start with the flow specification of a starting component, followed by a sequence of connections and other component flow specifications, ending with a component flow specification. The three types of flow declarations are summarized in Table 11-1.

You use flows to analyze important system characteristics. For example, you define a flow within a speed control system. This flow may originate at a speed sensor, traverse through the control elements of the system, and terminate at a throttle actuator. The total time between sending a speed value and the actuation of the throttle in response to that value can be analyzed. You identify the elements participating in the flow using flow declarations, assigning latency times to each of those elements using property associations, and calculating the total time for the complete flow path. The analysis compares this calculated time with the required time for the path from the sensor and the actuator.

11.1.1 Declaring Flow Specifications

Flow specifications are contained within the **flows** section of a component type declaration and establish a component's role in a flow. They identify the component as a flow source, flow sink, or flow path through the interfaces (component features) of the component. A flow source is

Table 11-1: *Flow Declarations*

Flow Declaration	Description
Flow Specifications	Identify the role (sink, source, or path) and external aspects of a flow through a component
Flow Implementations	Elaborate flow specifications through subcomponents
End-to-End Flows	Define a complete flow path from a starting component to a final component

associated with a component in which the flow originates and identifies the component feature (e.g., port, feature group, or parameter) through which the flow emerges. Similarly, a flow sink is associated with a component in which a flow terminates and identifies the component feature through which the flow enters. A flow path declares that a flow passes through a component from an incoming to an outgoing feature. Table 11-2 summarizes flow specifications and shows their graphical representations.

The following box shows the template for the three flow specification declarations (flow source, flow sink, flow path). Flows are named elements. They are associated with a component such that flows originate from, terminate, or pass through a component via features declared within the flow specifications for the component.

```
name : flow source  <exit feature> [{properties}];
name : flow sink  <entry feature> [{properties}];
name : flow path  <entry feature> -> <exit feature>
     [{properties}];
```

Listing 11-1 shows a partial specification for a simplified cruise control system with flow source, flow sink, and flow path specifications within component type declarations. The lower portion of the table includes a graphical representation of the flow specifications as well as the port connections between the components. Notice that the flow path *brake_ flow* through the system component *cruise_control* has an **out event data port** as its origin and an **in data port** as its termination feature. Since a flow is abstract, it does not need to involve a single port type. A mix of port types is permitted (e.g., events, data, event data). Even in the case of a flow involving only data or event data ports, the data type associated with the ports can be different. However, the flow must be consistent with the direction of the information flowing through the features.

Table 11-2: *Flow Specifications*

Flow Specification	Description
⊢ Flow source	Defines a flow that originates within a component and emerges from the component through an outgoing feature
⊣ Flow sink	Defines a flow that terminates within a component and enters the component through an incoming feature
→ Flow path	Defines a flow that enters a component through an incoming feature and emerges through an outgoing feature

In the case of data access, it must be consistent with write or read, while for ports it must be consistent with the port direction.

A component may have multiple flow specifications that define alternative roles for that component. This may include multiple flows through a single port as shown in Listing 11-1 for the event data port *brake_event*. Notice also that the flow paths *brake_flow* and *control_flow* in the system component *cruise_control* define two separate flow paths using the same incoming and outgoing ports. In the device *brake_pedal*, two flow sources are defined allowing the device to be a source for two distinct flows.

Listing 11-1: *Flow Declarations Within a Component Type Declaration*

```
device brake_pedal
features
    brake_event: out event data port float_type;
flows
    Flow1: flow source brake_event;
    Flow2: flow source brake_event;
end brake_pedal;
--
system cruise_control
features
    brake_event: in event data port;
    throttle_setting: out data port float_type;
flows
    brake_flow: flow path brake_event -> throttle_setting ;
     control_flow: flow path brake_event -> throttle_setting ;
end cruise_control;
--
device throttle_actuator
features
    throttle_setting: in data port float_type;
flows
    Flow1: flow sink throttle_setting;
end throttle_actuator;
```

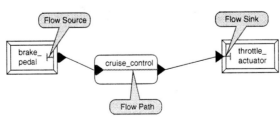

11.1.2 Declaring Flow Implementations

Flow implementations elaborate flow specifications of a component into a flow path through its subcomponents. Flow implementations identify the entry and exit features, connections, and subcomponents involved in a flow specification. For example, a flow source implementation for a component identifies the subcomponent from which the flow originates, all of the intervening subcomponents and connections that are involved in the flow within that component, and the feature through which the flow exits the component. A flow implementation may refer to only the entry and exit feature if no subcomponent exists or is involved in the flow.

Flow implementations are declared within the **flows** section of a component implementation. Patterns for each of the flow declarations are shown in the following box:

```
name : flow source  <source path sequence> -> <exit feature.
[{properties}];
name : flow sink  <entry feature> -> <sink path sequence>
[{properties}];
name : flow path  <entry feature> -> <flow path sequence>  ->
    <exit feature> [{properties}];
```

The name used for a flow implementation must identify one of the flow specifications of the component. Path sequences are alternating connection names and dot-separated subcomponent/flow specification name pairs that are separated by the connection symbol (->) that start and end with the entry and exit features of the corresponding flow specification. The path sequences for flow sources, flow sinks, and flow paths are shown in Table 11-3.

Entry or exit features identify the external feature of a component through which a flow enters or exits that component. Property associations enclosed in brackets {properties} can be included in the declaration.

In Listing 11-2, you can see an example of a flow source implementation. A flow source *disengage_flow* is declared in the system type *panel_system*. Its corresponding flow implementation declaration is contained in the **flows** section of the implementation *panel_system.impl*. In this declaration, the flow originates within the system *panel_switches* (from the flow source *disengage_flow*); traverses the connection *C1*; passes through the system *panel_control* via the flow path *disengage_flow_path*; traverses the connection *C2*; and exits through the port

Table 11-3: *Three Types of Flow Implementations*

Path Sequence	Description
Source	The flow implementation of a flow source specification starts with a flow source of a subcomponent as the flow origin, continues through zero or more subcomponent flow paths via connections, and ends with the exit feature of the flow source specification (e.g., *orig.flow0a -> Conn01 -> sub.flow0b -> Conn02->exitport*).
Sink	The flow implementation of a flow sink specification starts with the entry feature of the flow sink specification, continues through zero or more subcomponent flow paths via connections, and ends with a flow sink of a subcomponent as the flow destination (e.g., *entryport -> Conn01 -> dest.flow0b*).
Flow	The flow implementation of a flow path specification starts with the entry feature of the flow path specification, continues through zero or more subcomponent flow paths via connections, and ends with the exit feature of the flow path specification (e.g., *entryport -> Conn01 -> sub1.Path0a -> Conn02 -> sub2.Path0b -> Conn03 -> exitport*).

disengage_out. This can be seen graphically in the lower portion of Listing 11-2. In the graphic, the flow arrows are displaced slightly from the actual flow path.

Listing 11-2: *A Flow Source Implementation*

```
system panel_system
features
disengage_out: out event data port;
flows
disengage_flow: flow source disengage_out;
end panel_system;

system panel_switches
features
disengage_out: out event data port;
flows
disengage_flow: flow source disengage_out;
end panel_switches;

system panel_control
features
disengage_in: in event data port;
disengage_out: out event data port;
flows
disengage_flow_path: flow path disengage_in -> disengage_out;
end panel_control;
```

```
system implementation panel_system.impl
subcomponents
panel_switches: system panel_switches;
panel_control: system panel_control;
connections
C1: port panel_switches.disengage_out ->
    panel_control.disengage_in;
C2: port panel_control.disengage_out -> disengage_out;
flows
disengage_flow: flow source panel_switches.disengage_flow ->
    C1 -> panel_control.disengage_flow_path ->
    C2 -> disengage_out;
end panel_system.impl;
```

Listing 11-3 shows a flow sink implementation. In this example, the flow specification *e_flow_prop* defined in the type is detailed in the flow sink implementation declaration within the component implementation *PBA_emergency.impl*. The flow sink enters through the event data port *stop_prop* of the implementation, traverses connection *C1*, passes through the flow path subcomponent *emergency_control* (as flow path *e_flow_prop*), traverses the connection *C2*, and terminates in the device subcomponent *prop_lock* (as flow sink *e_flow_prop*). Notice that the nature of the flow changes from event data to event.

In this example the flow specification name is the same for different component types. However, the flow name *e_stop* is not permitted for the flow sink in the device type *propeller_lock*, since the device in event port is *e_stop*. This is allowed because flow specification names are local to a component type. This means that they cannot conflict with names of features or modes declared in the same component type.

Listing 11-3: *Example Flow Sink Implementation*

```
device propeller_lock
  features
    e_stop: in event port;
  flows
      e_flow_prop: flow sink e_stop;
end propeller_lock;
```

```
system emer_control
  features
    stop_prop: in event data port;
    e_stop_prop: out event port;
  flows
      e_flow_prop: flow path stop_prop -> e_stop_prop;
end emer_control;

system PBA_emergency
  features
    stop_prop: in event data port;
  flows
  e_flow_prop: flow sink stop_prop;
end PBA_emergency;

system implementation PBA_emergency.impl
  subcomponents
    emergency_control: system emer_control;
    prop_lock: device propeller_lock;
  connections
  C1: port stop_prop -> emergency_control.stop_prop;
  C2: port emergency_control.e_stop_prop -> prop_lock.e_stop;
  flows
      e_flow_prop: flow sink stop_prop -> C1 ->
emergency_control.e_flow_prop -> C2 -> prop_lock.e_flow_prop;
end PBA_emergency.impl;
```

Listing 11-4 shows the flow implementation declarations for the flow path *brake_flow* in the type declaration *cruise_control* of Listing 11-1. This flow path implementation originates at the *brake_event* event data port; traverses the connection *C1*; passes through the process component *data_in* via the flow path *interface_flow1*; traverses the connection *C3*; passes through the process component *control_laws* via the flow path *control_flow1*; traverses the connection *C5*; and exits through the data port *throttle_setting*. Notice that the nature of the data within the flow changes and involves event data ports as well as data ports. A graphical representation of the flow path is shown in the lower portion of Listing 11-4.

Listing 11-4: *Examples of Detailing a Flow Path Through a Component*

```
system implementation cruise_control.impl
subcomponents
data_in: process interface;
control_laws: process control;
connections
C1: port brake_event -> data_in.brake_event;
C3: port data_in.out_port -> control_laws.in_port;
C5: port control_laws.out_port -> throttle_setting;
flows
brake_flow: flow path brake_event -> C1 ->
    data_in.interface_flow1 -> C3 ->
    control_laws.control_flow1 -> C5 -> throttle_setting;
end cruise_control.impl;
--
process interface
features
brake_event: in event data port ;
out_port: out data port float_type;
flows
interface_flow1: flow path brake_event -> out_port;
end interface;
--
process control
features
in_port: in data port float_type;
out_port: out data port float_type;
flows
control_flow1: flow path in_port -> out_port;
end control;
```

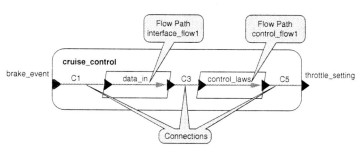

11.1.3 Declaring End-to-End Flows

An end-to-end flow details a path that originates at one component and terminates at another. This flow may traverse multiple components and their intervening connections. While end-to-end flows nominally involve a flow source component and a flow sink component, this need not be the case. Components with defined flow paths may originate or terminate an end-to-end flow. Note the contrast with detailing a flow path within a component, in which you define a path from one external feature of a component to another.

End-to-end flows are defined in the **flows** section of a component implementation declaration. The template for an end-to-end flow declaration is shown in the following box:

```
name :  end to end flow  <origin> → < path sequence> →
<destination>  [{properties}]  [in modes];
```

The origin is the subcomponent name followed by the name of its flow specification separated by a dot (e.g., *brake.flow1*). A path sequence consists of alternating connection names and dot-separated subcomponent/ flow specification name pairs that are separated by the connection symbol (→). The sequence begins with a connection name and ends with a connection name—possibly consisting of a single connection name. The destination consists of the destination subcomponent name followed by the name of its flow specification separated by a dot (e.g., *actuator.flow1*).

End to end flows can have property associations. They can represent desired properties of end-to-end flows or properties that have been calculated as part of an end-to-end flow analysis. The optional "in modes" statement associates the connection with specific modes or mode transitions.

You can leave off the flow specification for a component in a path sequence (e.g., if the component type of the subcomponent is missing flow specifications). In this case, a specification is automatically derived from the destination port of the preceding connection to the source port of the succeeding connection.

You can compose an end-to-end flow from other end-to-end flows. For example, you can define end-to-end flows E01 from A to B, E02 from C to D, and E03 from E to F. You can then define end-to-end flows e01 → conn → e02 and e01→ conn → e03 that represent two flows that branch at B. While it is not necessary, in this case, C is used as a common point to both flows.

The partial specification in Listing 11-5 illustrates an end-to-end flow *brake_flow* within the system implementation *complete.impl*. The flow originates at the device *brake_pedal* via flow source *Flow1*; traverses the connection *C1*; passes through the component *cruise_control* via the flow path *brake_flow*; traverses the connection *C2*; and terminates at the component *throttle_actuator* via the flow sink *Flow1*. A graphical representation of the end-to-end flow is shown in the lower portion of Listing 11-5.

Listing 11-5: *An End-to-End Flow*

```
system implementation complete.impl
subcomponents
brake_pedal: device brake_pedal;
cruise_control: system cruise_control;
throttle_actuator: device throttle_actuator;
connections
C1: port brake_pedal.brake_event -> cruise_control.brake_event;
C2: port cruise_control.throttle_setting -> throttle_actuator.
throttle_setting;
flows
brake_flow: end to end flow brake_pedal.Flow1 -> C1 -> cruise_control.
brake_flow -> C2 -> throttle_actuator.Flow1;
end complete.impl;
--
device brake_pedal
features
brake_event: out event data port;
flows
Flow1: flow source brake_event;
end brake_pedal;
--
system cruise_control
features
brake_event: in event data port;
throttle_setting: out data port float_type;
flows
brake_flow: flow path brake_event -> throttle_setting;
end cruise_control;
--
device throttle_actuator
features
throttle_setting: in data port float_type;
flows
Flow1: flow sink throttle_setting;
end throttle_actuator;
--
data float_type
end float_type;
```

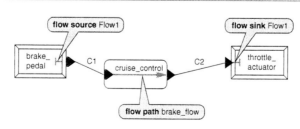

11.1.4 Working with End-to-End Flows

You can represent any logical flow through a system. The flow may start out as a flow through a data port of one data type, go through several event data ports and a write followed by a read data access, and end with an event port. You define the flow incrementally by defining flow specifications for each of the components and then elaborate each flow specification in the component implementation. This allows you to define high-level models (e.g., the top-level of a system of systems) and perform a first end-to-end flow analysis without further details of the system implementations. In other words, you perform a specification-based analysis.

When you have elaborated a system component through its component implementation declaration and added flow implementations, you can perform an analysis in two ways. In the first way, you can follow the flow implementation path and compare the results against the flow specification property. This only works if the analysis is compositional (i.e., can be performed incrementally down the component hierarchy). Alternatively, you can perform an end-to-end analysis on the instance model. In this case, the instance model generator expands the end-to-end flow instance down the component hierarchy to involve the leaf components as shown in Figure 11-1.

11.2 Binding Software to Hardware

An AADL model of an embedded system consists of a software application modeled as threads, processes, data components, connections, and remote subprogram calls, and a computer system modeled as processors, memory, buses, as well as virtual processors and virtual buses. In order to execute the software on the computer hardware you must

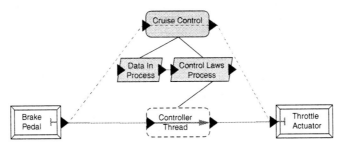

Figure 11-1: *End-to-End Flow Instance*

specify how the software is bound to the hardware. This is often referred to as deployment of the software on hardware. Without such a deployment you cannot perform analysis of operational properties such as performance, fault tolerance, safety criticality, and so on through static analysis and simulation or the generation of a build configuration.

First, we will show you how you use properties to declare such bindings and how you can represent deployment configurations. Then, we will describe to you the different bindings you can specify.

11.2.1 Declaring Bindings with Properties

AADL has a set of predeclared properties that you can use to specify bindings. One set of properties allows you (or a tool) to record the actual binding (e.g., the actual processor binding via the *Actual_Processor_Binding* property). This property takes as value the reference to the processor that a thread is bound to, and the property is associated with the thread. You do this by declaring the property association in a high-level component that contains both the software thread and the processor hardware, often the top-level system.

Listing 11-6 shows examples of actual binding declarations. The *Actual_Processor_Binding* declaration refers to *Speed_Processor* in the *Compute_Platform*. The property is applied to the *Speed_Control_Thread* in the *Speed_Control_Process*, which itself is contained in the application *speed_controller* (the component type and implementation declarations of the speed control process and thread are not shown).

In this example, we have placed all application software components into the *speed_controller* system and all computing hardware components in the *platform.singlePC* system component. The top-level system *speed_control_system.complete* is then defined to consist of the application software, the computer platform, and the sensor and actuator. Given this top-level system implementation declaration, you can then find the system implementation *speed_control_system.deployment1* as an extension of *speed_control_system_complete*. This extension contains only the binding properties for the system, effectively representing a deployment configuration.

Listing 11-6: *A Deployment Configuration*

```
system platform
  features
    marinebus: provides bus access standard.marine_certified;
end platform;
```

continues

```
system implementation platform.singlePC
  subcomponents
    speed_processor: processor pc.marine_certified;
    marine_bus: bus standard.marine_certified;
    DRAM: memory DRAM.marine_certified;
  connections
    bus access marine_bus <-> marinebus;
    bus access marine_bus <-> speed_processor.marinebus;
    bus access marine_bus <-> DRAM.marinebus;
end platform.singlePC;

system implementation speed_control_system.complete
  subcomponents
    speed_sensor: device sensor.speed;
    throttle_actuator: device actuator.throttle;
    speed_controller: system control.speed;
    compute_platform: system platform.singlePC;
  connections
    SC01: port speed_sensor.sensor_data ->
              speed_controller.sensor_data;
    CC01: port speed_controller.command_data ->
              throttle_actuator.command_data;
    PhConn1: bus access compute_platform.marinebus <->
                  speed_sensor.marinebus;
    PhConn2: bus access compute_platform.marinebus <->
                  throttle_actuator.marinebus;
  properties
    Allowed_Processor_Binding =>
      (reference (compute_platform.speed_processor));
    Allowed_Memory_Binding_Class => (classifier(DRAM));
end speed_control_system.complete;

-- possibly placed in a separate package
system implementation speed_control_system.deployment1
extends speed_control_system.complete
properties
Actual_Processor_Binding => (reference (speed_processor))
  applies to
  speed_controller.speed_control_process.speed_control_thread;
Actual_Connection_Binding => (reference (marine_bus))
  applies to SC01;
Actual_Memory_Binding => (reference (DRAM)) applies to
    speed_controller.speed_control_process;
end speed_control_system.deployment1;
```

In addition to the actual bindings, you can specify constraints on the bindings. For example, you can restrict the memory binding to memory of a certain type. This is shown in Listing 11-6 for the *Allowed_Memory_Binding_class* property. Similarly, you can indicate that the binding is constrained to a specific set of platform component (e.g., to specific processors by using the *Allowed_Processor_Binding* property).

These constraint properties are *inherited* (i.e., they apply to all components in the *speed_control_system.complete*) for which the properties are declared. You can also specify that two application software components or connections must be co-located or must not be co-located. You use this capability to ensure that duplicate software components are placed on redundant hardware components in fault tolerant system configurations.

11.2.2 Processor Bindings

The processor binding property *Actual_Processor_Binding* specifies which processor is responsible for executing a particular application thread. You can specify that a thread is bound to one processor or a list of processors, or to a system that contains processors. Using the processor list or system you can for example model that a thread can be executed on either core of a dual-core processor. You can apply the property to a single thread, to a list of threads, or to the enclosing thread group or process. For thread groups and processes, this means that all contained threads are bound to the specified processor(s). You can also use the processor binding properties to specify binding of threads to a virtual processor, and the binding of virtual processors to other virtual processors and to processors. In other words, you can use the virtual processor to represent virtual machines or hierarchies of schedulers. By binding an application thread to a specific virtual processor, you are choosing a particular scheduler. This scheduler itself is then bound to an actual processor. For example, you can have a processor with a rate-monotonic scheduler and a virtual processor that represents a sporadic server for event-driven threads. The sporadic server virtual processor is bound to the rate-monotonic processor to indicate that it takes one task slot, competing with periodic threads. Similarly, you can use virtual processors to represent rate group optimization. Rate group optimization means that a collection of application threads with the same period are executed by a single operating system task that calls the source code of each application thread in sequence. Each of these operating system tasks is a virtual processor that schedules the execution of application threads—all executing at the same rate on the same processor.

11.2.3 Memory Bindings

The memory binding property *Actual_Memory_Binding* specifies where the source code and the data are stored. You can specify such a binding for a system, process, thread group, and thread. You can also specify the binding for individual data components (e.g., global state variables within a process may be bound separately from the code of the process). You can also specify a separate binding for individual pieces of code (e.g., by binding subprograms and subprogram groups). Finally, you can specify memory bindings for ports (i.e., their location may be different from that of other data or code). As with processor bindings, memory bindings are inherited. For example, they can be declared for a process and they apply to all components inside the process unless those components have their own memory binding. The binding value can be a list of memory components or in the case of memory binding constraints, a list of memory classifiers.

11.2.4 Connection Bindings

The connection binding property *Actual_Connection_Binding*, specifies a routing of connections through hardware components. These can be port connections as well as data access or subprogram access connections. For example, if two application threads communicate via a port connection and they are bound to two different processors, the connection must be routed through a bus that physically connects the processors. In many cases, the connection routing can be inferred from the processor binding of threads. However, in some cases multiple physical paths may exist (e.g., two processors may be physically connected via dual-redundant buses). In such a case, it may be necessary to identify the particular bus over which the connection should be routed.

Connection binding can involve combinations of buses, processor, memory, devices, as well as virtual buses. You can use the virtual bus to indicate protocols to be involved in the connection. You can also indicate the need for a particular protocol and the fact that a protocol is provided by a processor or a bus through the *Provided_Virtual_Bus_Class* and *Requried_Virtual_Bus_Class* properties. Finally, you can indicate the need for a protocol by specifying a required quality of service level such as guaranteed delivery, ordered delivery, or secure delivery by using the *Required_Connection_Quality_Of_Service* and *Provided_Connection_Quality_Of_Service* properties. They can be used as constraints for a tool that determines actual bindings, or they can be used by a consistency checker to validate manually entered actual connection bindings.

11.2.5 Binding Remote Subprogram Calls

In Section 10.7, we introduced two ways of modeling remote subprogram calls. One of these ways is through subprogram access features and subprogram access connections (see Section 10.7.3). In that case, you can make use of the connection binding properties to specify a routing of the remote call in terms of the computer hardware. The second way of specifying a remote subprogram call is by identifying the thread that services the call through a call binding property (see Section 10.7.2). In this case, there is no connection through which you can specify the call routing in terms of the computer hardware. Instead, you have additional properties that allow you to specify this binding of the call, namely, *Actual_Subprogram_Call_Binding* and *Allowed_Subprogram_Call_Binding*. The use of the *Actual_Subprogam_Call_Binding* property is illustrated in Listing 11-7. It specifies a path from the processor *proc1*, which executes the caller code, via the bus *vme* to the processor *proc2*, which executes the thread with the remote subprogram.

Listing 11-7: *Subprogram Call Binding*

```
system implementation client_server_sys.impl
subcomponents
client_process: process client_process.impl;
server_process: process server_process.impl;
VME: bus VME;
RAM1: memory RAM;
RAM2: memory RAM;
Proc1: processor Intel;
Proc2: processor Intel;
connections
c1: bus access VME -> Proc1.Bus01;
c2: bus access VME -> Proc2.Bus01;
properties
Actual_Subprogram_Call =>
    reference(server_process.server_thread.service)
    applies to client_process.calling_thread.call_server;

Actual_Subprogram_Call_Binding =>
    (reference(Proc1), reference(vme), reference(Proc2))
  applies to client_process.calling_thread.call_server;

--
Actual_Memory_Binding => (reference (RAM1))
    applies to client_process;
Actual_Memory_Binding => (reference (RAM1))
    applies to server_process;
Actual_Processor_Binding => (reference (Proc1))
    applies to server_process.server_thread;
Actual_Processor_Binding => (reference (Proc2))
    applies to client_process.calling_thread;
end client_server_sys.impl;
```

continues

```
--
thread implementation calling.impl
calls   {
    call_server: subprogram service_it ;
        };
end calling.impl;
--
thread server_thread
features
service: provides subprogram access service_it;
end server_thread;
--
thread implementation server_thread.impl
end server_thread.impl;
--
subprogram service_it
end service_it;
--
```

Chapter 12

Organizing Models

In this chapter, you will learn about organizing your models. First, you assign unique identifiers within name scopes to model elements and use these names to uniquely reference them. Second, you use AADL packages to organize component type, component implementation, and feature group type declarations as well as annex libraries into component libraries. Third, you evolve classifiers by declaring them as incomplete classifier declarations (using **abstract**) and then refine them through **extends**. Finally, you can declare parameterized component templates (using **prototypes**) to specify architecture patterns and instantiate them with classifiers supplied as prototype actuals.

12.1 Naming and Referencing Model Elements

In AADL, all identifiers consist of one or more letters, digits, and the underscore ("_") symbol, starting with a letter. The identifier cannot have two consecutive underscores.

12.1.1 Naming and Referencing with Packages

Packages are named by a sequence of identifiers, separated by double colons ("::"). The package names must be unique with respect to other package names and property set names. An example is *Supplier1:: PowerTrain::ETC::Software*. You can find details about organizing models with packages in Section 12.2.

Packages provide a name scope for component types, component implementations, and feature group types. This means the names of these classifiers must be unique within a package, but the same classifier name can be used in different packages. In addition, any alias identifier for type references or package names—introduced by a *renames* (Section 12.2.3)—must be unique within a package. You reference these classifiers uniquely by qualifying them with the package name. Qualifying a classifier in the same package is optional. When you reference a classifier in another package that package must be listed in the **with** statement of the package with the reference (see Section 12.2.2 for details).

12.1.2 Naming and Referencing Classifiers

A single identifier names a component type or a feature group type. These names must be unique within the package within which they are declared. You reference the process type *MyProcess* by qualifying it with the package name (e.g., *Supplier1::PowerTrain::ETC::Software:: MyProcess*).

Component implementations are named by their component type identifier and an implementation identifier separated by a period ("."). The implementation identifier must be unique within the component type; this makes the component implementation names unique within a package. You reference the process implementation *MyProcess.secure* by qualifying it with the package name (e.g., *Supplier1::PowerTrain:: ETC::Software::MyProcess.secure*).

Components provide a name scope for model elements declared within them. This means that for component types (Section 5.3), the identifiers for prototypes, features, flow specifications, modes, and mode transitions must be unique within a component type, and for component implementations (Section 5.5), the identifiers of prototypes, subcomponents, connections, modes, mode transitions, call sequences, and calls must not conflict with each other nor the model elements declared in the component type. For example, you cannot have a port and a subcomponent with the same name.

Feature groups provide a name scope for feature group elements. This means that for feature group types (Section 10.4.1) the identifiers of features and prototypes must be unique within the feature group type. For example, a feature group type cannot have a port and a feature group with the same name.

12.1.3 References to Model Elements

Within a component implementation you reference features of subcomponents in connection declarations by the subcomponent identifier and the feature identifier separated by a period ("."), and you reference a feature of the component itself (e.g., the component whose implementation contains the connection declaration) simply by the feature identifier. In the case of a feature group, you can also reference an element of a feature group by the feature group identifier and the element feature identifier separated by a period ("."). In the case of access connections, you may also reference a subcomponent by itself. You can find examples of connection declarations in Chapter 10. You can reference subprogram and subprogram group access features in subprogram call declarations. You do so by the subprogram access feature identifier, or the subprogram group access feature identifier followed by (".") and the subprogram access feature within the subprogram group (Section 10.7).

Within contained property associations you can indicate the model element(s) to which the property value is assigned by a reference path in an **applies to** statement (Section 13.2.4). The reference path is a sequence of identifiers separated by ("."), where the first identifier must exist in the name scope of the location the property association is declared. When declared in the properties section of a component type, or feature group type, the path can refer to elements in the type. For example, you can associate a property value with a data port or an event port within a feature group. When declared in the properties section of a component implementation, the path can refer to elements in the type and in the implementation. For example, you can assign an implementation specific property value to a port, or you can assign a property value to a connection.

The reference path of the **applies to** statement in a contained property association of component implementations and subcomponents can also reference a model element further down the component hierarchy. In the case of a component implementation, the initial identifier(s) refer to subcomponents starting with the subcomponent of the component implementation. In the case of a contained property association declared with a subcomponent (Section 9.1.1), the first identifier must be a subcomponent of this subcomponent. In other words, the starting point of the reference path is the location at which the property association is declared.

For reference paths that are the value of a property association of type **reference** (Section 13.2.2), the starting point for resolving the reference is the location of the property association declaration.

12.1.4 Naming and Referencing with Property Sets

A single identifier names a property set. Unlike packages, their name cannot consist of multiple identifiers. Their name must be unique with respect to other property sets and packages.

Property sets provide a name scope for property definitions, property types, and property constants (Section 14.1). In other words, the identifier of a property definition, property type, or property constant must be unique within a property set. For example, you cannot have two properties with the name *SecurityLevel* in a single package.

References to these items are qualified by the property set name separated by ("::"). All references including those within the same property set must be qualified (e.g., *Security::SecurityLevel*). The only exceptions are properties, property types, and property constants predeclared in the standard [AS5506A] (e.g., the property *Period*). When you reference an item in a different property set, this property set must also be listed in the **with** statement of the package or property set containing the reference.

12.2 Organizing Models with Packages

You use AADL *packages* to organize the different parts of your AADL model. You do this the same way as you would use packages in Java or UML to organize Java classes and methods in Java or detailed design diagrams in UML. Packages contain component type, component implementation, feature group type, and annex library declarations. Packages have a *public* section and a *private* section. You can only reference the items in the public section from other packages.

Each package provides a separate name scope (i.e., two different packages can contain declarations with the same name). As mentioned before, the declared items are referenced by the package name and the item name.

Packages have a nested naming hierarchy. The complete path in the naming hierarchy uniquely identifies a package (e.g., *Supplier1:: PowerTrain::Engine::ETC::Software*). However, packages are not placed inside each other. Instead, you can store them in the file system or in a model repository as separate files or version-controlled units. Packages in different parts of the naming hierarchy can have the same name. The naming hierarchy does not restrict which package can use the public items of other packages. Instead, each package indicates through a **with** clause the set of other packages it intends to utilize.

You can use the nested package names to organize your component libraries and annex libraries. You can use naming conventions that you are familiar with from packages in other languages. If different teams or different organizations develop your system, you may want to include the name of the organization or team in the package name hierarchy. You may want to place hardware libraries under one name subtree, your physical system components under a second, and your software libraries under a third. For the software, you may want to distinguish between the runtime architecture in the form of threads and processes and the application program architecture in the form of data types and functions. You normally reflect the hierarchical structure of the runtime architecture including the task and communication architecture, the computer hardware architecture, and the physical system in the component hierarchy of the model. You are not required to reflect this hierarchy in the nested package name hierarchy, although it is common to group component declarations into packages to represent libraries and organize the libraries according to some aspects of the system hierarchy.

12.2.1 Declaring Packages

You can declare a package with a public section only, or with a public and a private section. You can also have two separate package declarations for the same package, one with only a public section and the other with only the private section. This option allows you to make a file containing a package available to others without including its private items. Elements in the public segment are accessible outside the package, whereas elements in the private sections are accessible only within the package. You cannot place a package declaration inside another package. Listing 12-1 shows the template for a package declaration. You can place classifier declarations in any order.

Listing 12-1: *Package Declaration Template*

```
package <package name>
  public
    <visibility declarations>
    <classifier declarations>
```

continues

```
   private
     <visibility declarations>
     <classifier declarations>
   [ properties
     <property associations> ]

end <package name>;
```

The package name is a sequence of identifiers separated by double colons ("::"). A package name looks like *primary_control_system::roll_axis::control_components*. This naming flexibility can be useful for packages that have been developed independently and have been assigned the same name. For example, consider two engineering teams working on a project, *team red* and *team blue*. Each team develops a package with the name *sensor_control*. These package names can be made unique by prefixing them with the team name (e.g., *team_red::sensor_control* and *team_blue::sensor_control*).

A package must contain a public or private section and may contain both. You can declare packages whose public or private section is initially empty. If the optional **properties** section is included, at least one entry or the reserved word **none** and a semicolon must appear within that section. The property associations declared in the properties section apply to the package itself not to its contents.

The public or private section consists of visibility declarations followed by classifier declarations. Visibility declarations are **with** statements and **renames** statements. Classifier declarations are component type, component implementation, or feature group type declarations as well as annex library declarations.

The **with** statement allows you to specify the other packages that are visible to the content of a package. In other words, you can explicitly declare which packages you intend to make use of in a package and the AADL compiler will enforce these visibility rules. This is useful when you want to manage the degree of dependencies between packages, or limit how a person responsible for a package can make use of other packages.

The **renames** statement allows you to introduce a local short name (alias) for references to classifiers in other packages, as these referenced packages and classifiers can have long names.

Listing 12-2 shows an example package declaration. This package has both public and private sections that include component declarations and a properties section. The data component *new_format* declared in the private section of the package cannot be accessed from outside.

However, the data component *display_data* can be, since it is declared in the public section of the package.

Listing 12-2: *Example Package Declaration*

```
package display_dynamics_set

-- Elements accessible from outside the package are listed
-- following the key word public
public
process compress_data
features
display_data_input: in data port display_data;
formatted_data: out data port;
data_error: out event port;
end compress_data;

process implementation compress_data.impl
end compress_data.impl;

data display_data
end display_data;
-- Elements accessible only inside the package are listed
-- following the key word private
private

data new_format
end new_format;

properties
none;

end display_dynamics_set;
```

12.2.2 Referencing Elements in Packages

When you reference a classifier in your own package, you can do so by just naming the classifier. In your own public section, you can reference only classifiers in the public section. In your own private section, you can reference classifiers in both the public and private section.

When you want to reference classifiers in the public section of other packages you must preface the classifier name with the full package name followed by a double colon and the classifier name. Component types or implementations placed in the private section of a package cannot be referenced from outside the package. For example, the process implementation *compress_data.impl* contained in the public segment of the package *display_dynamics_set* shown in Listing 12-2, would be referenced from outside the package as *display_dynamics_set::compress_data.impl*, as shown in Listing 12-3.

Listing 12-3: *Referencing an Element Contained Within Another Package*

```
system implementation display_management.impl
subcomponents
compress_data: process display_dynamics_set::compress_data.impl;
….
end display_management.impl;
```

To reference classifiers located in other packages, you must have declared the name of the package being referenced in a **with** declaration. You can list multiple package names in a single **with** declaration, or you can list each in a separate **with** declaration. The **with** declaration may be included in either public or private sections or both. Examples are shown in Listing 12-4, where the **with** statements in the public section identify two packages *control* and *display_dynamics_set::display_data* for which the contents of their public sections can be accessed from the public section of the package *local_display_set*. The declaration for in data port *data_input* references the data type *core_data* contained in the package *display_dynamics_set::display_data*. The subcomponent *compress_data* is an instance of the process implementation *compress_data. impl* that is contained in the package *control*.

In the private section, the package *error_set::standard_errors* is declared in a **with** declaration and the data component *error_format* within that package is referenced in the out data port declaration *err_data*. The data type declaration of *basic_format* in the private section references the data type *b_formats* located in the package *display_dynamics_set::display_data*. It can do so because the package is already listed in the **with** declarations of the public section.

Notice that the package name is optional in referencing elements within the same package, even if they are between the private and public section. The package is used in referencing *scale_data.impl* for the process subcomponent *scale_data*. However, it is not included for the redundant subcomponent *scale_data_redun*.

Listing 12-4: *Example Package That Includes a with Statement*

```
package local_display_set

public
   with control;
   with display_dynamics_set::display_data;
```

```
  process convert
  features
    data_input: in data port
      display_dynamics_set::display_data::core_data;
    formatted_data: out data port;
    data_error: out event port;
  end convert;

  process implementation scale_data.impl
      extends control::scale_data.impl
  end scale_data.impl;

  system display_management
  end display_management;

  system implementation display_management.impl
  subcomponents
    compress_data: process control::compress_data.impl;
    scale_data: process local_display_set::scale_data.impl;
    scale_data_redun: process scale_data.impl;
  end display_management.impl;

private
  with error_set::standard_errors;

  process manager
  features
    err_data: out data port
      error_set::standard_errors::error_format;
  end manager;

  data basic_format extends
      display_dynamics_set::display_data::b_formats;
  end basic_format;

end local_display_set;
```

12.2.3 Aliases for Packages and Type References

You can introduce an alternative short name or alias for a package, a component type, or a feature group type that you reference. You define these aliases using a **renames** declaration. The alias identifier must be unique within the package in which the alias is declared.

Examples of package renaming declarations are included in the package *local_display_set* shown in Listing 12-5. This example is a modification of the specification found in Listing 12-4. It contains four types of alias declarations and their use.

First, the alias *e_pkg* is defined for the package *error_set::standard_errors* in the private section using the keywords **renames package**. The alias is then used in the **extends** declaration of the *basic_format* data component type. This renaming capability is useful in introducing shorter names for lengthy package names.

Second, you can declare an alias whose name is different from the original type classifier name (i.e., that of a component type or feature group type) using a **renames** declaration. The alias *core_d* declares a short name for a data component type in another package. This alias is then used as classifier in the *data_input* port of process *convert*. You can also reference a component implementation of this component type by naming the alias followed by a dot (".") and the implementation identifier—just as you would reference a component implementation in the same package. This capability allows you to use an alternative or shorter name for type classifiers referenced in the public section of external packages.

Third, you can use the name of a component type from another package as the alias identifier. To do this, use a **renames** declaration without a new identifier, as shown in the second renames declaration in the package *local_display_set* in Listing 12-5. This declaration references the data component type *basic_data* that is contained in the package *display_dynamics_set::display_data*. As a result, the reference to *basic_data* in the out data port declaration of *formatted_data* is to *basic_ data* contained in the package *display_dynamics_set::display_data*. The *basic_data* alias is also referenced by the data implementation declaration *basic_data.impl*. This illustrates how you can declare a component type in one package and then place its component implementations in a separate package.

Listing 12-5: *Examples of Aliases*

```
package local_display_set

public
  with control;
  with display_dynamics_set::display_data;
  core_d renames display_dynamics_set::display_data::core_data;
  renames display_dynamics_set::basic_data::basic_data;
  renames control::all;

process convert
  features
  data_input: in data port core_d;
  formatted_data: out data port basic_data;
  data_error: out event port;
  end convert;
```

```
system implementation display_management.impl
subcomponents
compress_it: process compress_data.impl;
scale_data: process scale_data.impl;
end display_management.impl;

data implementation basic_data.impl
end basic_data.impl;

private
with error_set::standard_errors;
e_pkg: renames package error_set::standard_errors;
data basic_format extends e_pkg::b_formats;
end basic_format;

end local_display_set;
```

Finally, you can make all public classifiers of a package available as aliases using their original name by declaring **renames** with a package name followed by a double colon and the reserved word **all**. This is illustrated with the third **renames** declaration in the package *control*. The subcomponent *compress_it* then references the implementation *compress_data.impl* of component type *compress_data*. Remember that the alias names must be unique in the package containing the alias declaration. For example, you cannot declare a data component type *compress_data* within the package *local_display_set*, since it would conflict with the type *compress_data* declared in the package *control* and has been introduced as alias. Nor can you declare a component type *basic_data* since it would conflict with the type in the package *display_dynamics_set::display_data*.

12.3 Evolving Models by Classifier Refinement

Using an extension declaration, you can define new classifiers (component types, component implementations, and feature group types) by extending an existing classifier. The new classifier inherits all of the characteristics of another classifier. You use extensions to add new elements, refine existing elements inherited from the extended classifier, refine them to be mode specific, assign new property values to the new classifiers or any of its elements, and to refine the **abstract** category into one of the concrete categories. Additions can include new features, flow specifications, modes, and annex subclauses for component types, subcomponents, connections, modes, mode transitions, call sequences, flows, and annex subclauses for component implementations, and features for feature group types. In refinements of existing elements such as features and subcomponents, you add missing classifier references, add an implementation to an existing type reference, substitute an existing classifier reference with another that satisfies specific substitution rules, indicate that the element is mode specific, or assign a property to an existing element.

You use classifier extension to declare a family of component types and variants of component implementations. You also use classifier extension to refine a collection of partially declared classifiers that represents a reference architecture into an instance of this architecture [NASA]. When you extend a partially declared classifier, you apply a component pattern by filling in the missing parts. In particular, you may want to make use of prototypes to explicitly parameterize such classifier patterns (see Section 12.4). You also use classifier extension to configure a complete model with property values for its model elements by declaring them as a collection of contained property associations (see Section 13.2.4).

12.3.1 Declaring Classifier Extensions

Classifier extensions are declared by referencing an existing classifier with an **extends** clause. The classifier extension can add new elements to the classifier being extended and it can refine elements of the original classifier. Table 12-1 shows the system extension *controller_low_band* that extends *controller_basic* by adding two new features. The system type *controller_low_band* consists of five data ports, three inherited from *controller_basic* and the two additional ports declared as part of the extension.

Table 12-1: *Basic Extensions with Additions*

`system controller_basic`
`features`
`vehicle_speed: in data port;`
`set_point: in data port;`
`command_out: out data port;`
`properties`
`Period => 50 ms;`
`end controller_basic;`
	`system controller_low_band`
	`extends controller_basic`
	`features`
	`engine_speed: in data port;`
	`acceleration_rate: in data port;`
	`end controller_low_band;`

12.3.2 Declaring Model Element Refinements

An extended classifier can also refine an element of the original classifier by redeclaring the element with the keywords **refined to** added, as shown in the following box for a subcomponent refinement.

```
Thread1 : refined to thread in modes ( model );
```

You can refine features in component types and feature group types, and you can refine subcomponents in component implementations to add or substitute classifier references. By default, you can add a component type if there was none, add a component implementation if there was no classifier or the identical component type, or you can replace a component implementation with another of the same component type. Other refinement substitution rules are described in Section 12.3.3. You can also refine flow specifications, end-to-end flows, subcomponents, and connections to be mode-specific by adding an **in modes** clause. Finally, you can use the refinement to add property associations to component type features, flow specifications, and modes, feature group type features, as well as subcomponents, connections, modes, mode transitions, and calls in component implementations.

In a refinement declaration the classifier reference, in mode clause, and property association in curly brackets are optional. In other words, you only need to declare those clauses that change. In the case of connection refinements, you do not redeclare the source and destination of the connection.

In the example of Table 12-2, the system type for a basic controller (shown on the left) is extended to define a high fidelity system type with two additional input ports and the refinement of the port declarations of the *vehicle speed, engine_speed,* and *acceleration_rate* to add the data component type as classifier reference. The *Period* property value 20ms applies to instances of the system type *controller_high_fidelity* and any of its implementations (unless they have their own *Period* value assigned), while the *Period* value of the 50ms value applies to all instances of *controller_basic* and its implementations.

When you extend a component type, the component type extension does not automatically inherit the implementations of the original. Instead, you must explicitly declare each implementation (e.g., as extension of one of the implementations of the original component type) possibly without adding or refining elements.

Table 12-2: *Extending a Component Type Declaration*

`system controller_basic`
`features`
`vehicle_speed: in data port;`
  `set_point: in data port;`
  `command_out: out data port;`
`properties`
  `Period => 50 ms;`
`end controller_basic;`
	`system controller_high_fidelity`
	`extends controller_basic`
	`features`
	  `vehicle_speed: refined to in data port`
	    `HF_speed_data;`
	  `engine_speed: in data port basic_data;`
	  `acceleration_rate: in data port accel_data;`
	`properties`
	    `Period => 20 ms;`
	`end controller_high_fidelity;`
	
	`data HF_speed_data`
	`end HF_speed_data;`
	
	`data basic_data`
	`end basic_data;`
	
	`data accel_data`
	`end accel_data;`

You can declare a component implementation extension by referencing a component implementation of the same component type, or a component implementation whose component type is the origin of the extension's component type. This is illustrated in Listing 12-6, where the extended process implementation *control_high_fidelity.HF_control* is of type *control_high_fidelity* in Table 12-2, which is an extension of the process type *control*. It extends *control.basic*, which is an implementation of *control_high_fidelity*'s "ancestor" type. In the process implementation *control.basic*, only the thread type is referenced in subcomponent declarations. Since data ports are defined in the type, connections can be declared in the implementation *control.basic*. This implementation can serve as a pattern for subsequent extensions. In the implementation extension *control_high_fidelity.HF_control*, the thread subcomponents are refined to reference specific thread implementations. A binding property is added in the connection refinement for the connection *DC1*, while a binding property is added to connection *DC2* by a contained property association in the properties section of *control_high_fidelity. HF_control*. This illustrates the two ways in which you can add a property value to an existing model element: by model element refinement, and by contained property association.

Listing 12-6: *Examples of Implementation Extensions*

```
process implementation control.basic
subcomponents
  scale_speed_data: thread read_data;
  speed_control_laws: thread control_laws;
connections
  DC1: port sensor_data -> scale_speed_data.sensor_data;
  DC2: port scale_speed_data.proc_data ->
    speed_control_laws.proc_data;
  DC3: port speed_control_laws.cmd -> command_data
    {Allowed_Connection_Binding_Class =>
      (classifier(high_perf_bus.impl));};
  EC1: port disengage -> speed_control_laws.disengage;
  DC4: port set_speed -> speed_control_laws.set_speed;
end control.basic;

process implementation control_high_fidelity.HF_control
extends control.basic
subcomponents
  scale_speed_data: refined to thread read_data.HF;
  speed_control_laws: refined to thread control_laws.HF;
connections
  DC1: refined to port
    {Allowed_Connection_Binding_Class =>
    (classifier (high_perf_bus.impl));};
properties
  Allowed_Connection_Binding_Class =>
    (classifier (high_perf_bus.impl)) applies to DC2;
end control_high_fidelity.HF_control;
```

12.3.3 Classifier Substitution Rules for Refinements

You can alter the default rules for adding or substituting classifiers by using the *Classifier_Substitution_Rule* property within a classifier or subcomponent (or feature) declaration. The values for this property are *Classifier_Match, Type_Extension*, and *Signature_Match*.

The default value is *Classifier_Match*. As described in Section 12.3.1, this means that if the original had a component type, the component type must remain the same. In other words, if the original classifier is a component type, any implementation of that type is acceptable. If the original classifier is a component implementation, an alternate implementation of the same type is acceptable. If the original declaration is a feature group type without features, an extension of that feature group type can be substituted for the original in the refinement.

Type_Extension means that any component type that is an extension of the original component type being refined is an acceptable substitute.

In addition, any implementation whose type is identical or an extension of the original type can be used. Similarly, any feature group type that is an extension of the feature group being refined is an acceptable substitute.

Signature_Match means that the original type must have a subset of the features of the substitute type.

Consider the partial specification in Listing 12-7, where the process type *control* is extended to create the *hf_controller* and its implementation *control.impl* is extended to create *hf_controller.impl*. The property *Classifier_Substitution_Rule* is assigned the value *Type_Extension* in the type declaration *control*. Since it is defined as inherited property, the value *Type_Extension* for *Classifier_Substitution_Rule* is inherited by the implementations and extensions of *control*. In the extension declaration for *hf_controller*, the data port *input* is refined to a data implementation *hf_signal.basic*. This is a type extension of the data type *signal* originally referenced in *control*. This refinement is permitted since the value of *Classifier_Substitution_Rule* is *Type_Extension*, inherited from the type *control*.

The feature group *output_set* in the process type *control* is replaced by its extension *hf_output_set* in the type extension *hf_controller*. This substitution is permitted for either the *Classifier_Match* or the *Type_Extension* value of *Classifier_Substitution_Rule*, since the original feature group *output_set* does not contain any features. Note that the data port *output* is refined by simply adding the data implementation *cmd.data* reference. Again, this is permitted for both values of the *Classifier_Substitution_Rule*, since no classifier reference is included in the port declaration within *control*.

In Listing 12-7, the process implementation *hf_controller.impl* is an extension of *controller.impl*. Within this extension, the thread subcomponent *control_thread* is refined to *hf_control_laws.impl* from its original type *control_laws* declared in *controller.impl*. Again, this is possible since the value of the property *Classifier_Substitution_Rule* is *Type_Extension*, inherited from the ancestor type *control*.

Listing 12-7: *Refinement Using Extended Types*

```
process control
features
 input: in data port signal;
 output: feature group output_set;
properties
Classifier_Substitution_Rule => Type_Extension;
end control;
```

```
process hf_controller extends control
features
 input: refined to in data port hf_signal.basic;
 output: refined to feature group hf_output_set;
end hf_controller;

data signal
end signal;

data hf_signal extends signal
end hf_signal;

data implementation hf_signal.basic
end hf_signal.basic;

feature group output_set
end output_set;

feature group hf_output_set extends output_set
features
out_data: out data port cmd.basic;
error_data: out event data port error_data_set;
end hf_output_set;
--
process implementation control.impl
subcomponents
 input_thread: thread scale_data;
 control_thread: thread control_laws;
end control.impl;

process implementation hf_controller.impl extends control.impl
subcomponents
control_thread: refined to thread hf_control_laws.impl;
end hf_controller.impl;

thread control_laws
end control_laws;

thread implementation control_laws.impl
end control_laws.impl;

thread  hf_control_laws extends control_laws
end hf_control_laws;

thread implementation  hf_control_laws.impl
    extends control_laws.impl
end hf_control_laws.impl;
```

12.3.4 Refining the Category

It is possible to refine the component category of a component type or implementation from **abstract** to a specific runtime category as part of an **extends** declaration. You can refine the category of a subcomponent from **abstract** to one of the concrete categories as part of a **refined to** declaration. Finally, you can refine an abstract feature, declared as **feature**, to a specific port, data access, bus access, or subprogram access feature as part of a **refined to** declaration.

Listing 12-8 includes examples of category refinement where the implementation *monitor.generic* and its subcomponents are defined as abstract. These abstract components are extended into runtime components. For example, the system type *monitorRT* is an extension of the abstract type *monitor* and the system implementation *monitorRT.impl* is an extension of the abstract implementation *monitor.generic*. Within the extension *monitor.generic* the abstract subcomponents *acquire_data* and *detection* are refined into device instances.

Listing 12-8: *Category Substitution*

```
abstract monitor
end monitor;

abstract implementation monitor.generic
subcomponents
  acquire_data: abstract acquire;
  detection: abstract detect;
end monitor.generic;

abstract acquire
features
  input: in data port;
  output: out data port;
  bus_acc: requires bus access basic_bus;
end acquire;

abstract implementation acquire.basic
end acquire.basic;

abstract detect
features
  input: in data port;
  output: out data port;
  exhaustManifold: provides bus access basic_bus;
end detect;

abstract implementation detect.basic
end detect.basic;
```

```
-- runtime category substitution --

system monitorRT extends monitor
end monitorRT;

system implementation monitorRT.impl
   extends monitor.generic
subcomponents
    acquire_data: refined to device acquire.basic;
    detection: refined to device detect.basic;
end monitorRT.impl;
```

12.4 Prototypes as Classifier Parameters

You can declare a prototype as a parameter of a component type, component implementation, or feature group. Prototypes act as placeholders for classifiers (i.e., component type, component implementation, and feature group type). You can reference the prototype anywhere you would normally reference a classifier. You later supply a specific classifier, as actual, for the prototype when you reference a parameterized classifier (e.g., when refining the classifier using *extends* or when declaring a feature or subcomponent). You can also declare a prototype as an abstract feature to indicate that the specific feature, such as port or access, will be supplied later.

You use prototypes on classifier declarations when you want to parameterize a component with the classifiers to be used for its subcomponents or for data types on its features and be explicit about the classifier "parameters" to be supplied. This is useful when you want to represent reference architectures or configurable product line families. When you reference the same prototype in multiple places, you indicate that the same actual classifier should be used (e.g., as data type of a port, or as classifier of redundant subcomponents). In other words, you can declare parameterized classifier templates.

12.4.1 Declaring Prototypes

You declare prototypes as parameters for component types and implementations, and feature group types in the prototypes section. The parameters are declared to be classifiers, feature group types, or features using one of the forms shown in Listing 12-9, where non-bolded square brackets [ ] indicate optional entries. In using these declarations, you define a name for a prototype; identify what kind of prototype it is; and optionally constrain it with a specific classifier. The angular bracketed entry **\<category\>** indicates that a specific component category is to be included in its place.

Listing 12-9: *Prototype Declarations*

```
Declaring a Prototype for a Component Type or Implementation
name : <category> [<component classifier>] [ [] ][{properties}];

Declaring a Prototype for a Feature Group Type
name: feature group [<feature group type classifier>]
    [{properties}];

Declaring a Prototype for a Feature
name: [in|out] feature [<component classifier>][{properties}];
```

For component classifier prototypes, the prototype specifies at a minimum the component category of the expected classifier. If the category is *abstract,* any category is acceptable. If the category is one of the concrete categories, such as *thread,* the supplied prototype actual must be of that category.

The prototype declaration for component classifiers may include a reference to a component classifier. In this case, the component classifier supplied as the prototype actual must match this classifier according to prototype substitution rules. By default, the classifier must match. You use classifier match if you want to limit the prototype actual to be one of the implementations of the component type in the prototype declaration. The substitution rule may permit type extension. In that case, any component type or implementation that is an extension of the classifier in the prototype declaration is acceptable. The third option of the substitution rule allows for signature matching (i.e., the classifier supplied as prototype actual must have the same features as those of the classifier in the prototype declaration) but one is not required to be an extension of the other.

The optional square brackets for a component classifier prototype indicate that a list of classifiers is expected as prototype actual. These classifiers will then be used as actual classifiers for different elements of a subcomponent array. For example, you can declare a sensor array, where different elements are different implementations or variants of the same type. You can also use this to declare a redundancy pattern such as N-Version redundancy.

The prototype for feature group types indicates that a feature group type is expected as prototype actual. If a feature group type classifier is included in the prototype declaration, prototype substitution rules determine acceptable prototype actual. The rules for feature group types are the same as those for component types.

The prototype for features uses an abstract feature as the placeholder. This feature prototype is replaced by a concrete feature such as a port or access feature as prototype actual. The feature prototype may have a direction specified. In this case, the direction of the supplied feature must match. If the feature prototype has a classifier, then prototype substitution rules determine acceptable classifiers for the feature. For example, you can declare a prototype as an abstract feature with a specific data type. In this case, you limit the prototype actual to data ports, event data ports, or data access with the specific data type. In other words, the user of a classifier with this prototype can choose between a data port, an event data port, or data access, but cannot change the data type under the default substitution rule.

12.4.2 Using Prototypes

You can declare component classifier prototypes for component types, component implementations, and feature group types. You use them as placeholders by referencing the prototype in place of the classifier in feature and subcomponent declarations inside the component type, component implementation, or feature group type. For example, if you want to parameterize the data type of a port or a data access without refining the feature type, you make use of the component prototype by declaring a *data* component classifier as prototype. This prototype is then referenced in the port or data access declaration as shown in Listing 12-10. You can reference the same prototype in several feature or subcomponent declarations to indicate that they will have the same actual classifier. This is particularly useful when you define component patterns or reference architectures.

You can declare feature group type prototypes for component types and feature group types. You use them as placeholders by referencing the prototype in place of the feature group type in feature group declarations.

You can declare feature prototypes for component types and feature group types. You use them as placeholders by referencing the feature prototype in an abstract feature declaration after the keyword **feature**.

Example prototype declarations are shown in Listing 12-10. In the process type declaration called *basic*, the prototype *rd* is used as a placeholder for a data classifier for the input data port *input*. The feature group *cmd* is a prototype for the *output* feature group and the feature *cd* is an abstract feature prototype for the incoming feature *control*.

In the process implementation for the process type *basic*, the prototype *cl* is declared as a placeholder for a thread implementation of the type *control*. It is used in a declaration for the thread subcomponent *c_laws*. This prototype may be replaced with a specific implementation such as *control.PID* as prototype actual. The prototype *rt* is declared as an abstract component and is used in the subcomponent declaration for *r_data*. As a subcomponent of a process, the prototype actual for *r_data* is limited to a thread, thread group, data, subprogram, and subprogram group because a process can only contain subcomponents of those categories.

Listing 12-10: *Example Prototype Declarations*

```
process basic
prototypes
  rd: data;
  cmd: feature group;
  cd: feature;
features
  input: in data port rd;
  output: feature group cmd;
  control: in feature cd;
end basic;

process implementation basic.impl
prototypes
  cl: thread control;
  rt: abstract;
subcomponents
    c_laws: thread cl;
    r_data: abstract rt;
end basic.impl;
```

12.4.3 Providing Prototype Actuals

You can provide prototype actuals to a classifier with prototypes at the time you declare an extension of this classifier, or you can provide the prototype actuals at the time you reference such a classifier (e.g., in feature, feature group, or subcomponent declarations). You use the forms shown in Listing 12-11 to bind an actual classifier or feature as actual value to a prototype. The general form is a comma-separated list of one or more individual prototype bindings enclosed in parentheses. You may provide prototype actuals for a subset of the prototypes. In that case, you can then extend the resulting classifier by supplying prototype actuals for prototypes without actuals.

In the case of a component classifier prototype, you may supply a list of classifiers if you declared the prototype to expect such a list by [].

When you provide a prototype actual to a prototype in the classifier reference of a feature or subcomponent declaration, you can reference a prototype of the enclosing classifier (i.e., the classifier that contains such a declaration). This allows you to pass prototype actuals down the component hierarchy. In other words, you can parameterize the component hierarchy with classifiers across multiple levels the same way that you can parameterize it with property values by using contained property associations.

Listing 12-11: *Providing Prototype Actuals*

```
component prototype binding:
prototype_name => component prototype actual
prototype_name => (comma-separated list of prototype actual)
    component prototype actual:
    category component_classifier [prototype_bindings]
    category enclosing_component_classifier_prototype_name

feature group prototype binding:
prototype_name => feature group feature_group_classifier
prototype_name => feature group
  enclosing_feature_group_type_prototype_name

feature prototype binding:
prototype_name => <port description> component_classifier
prototype_name => <access description> component_classifier
prototype_name => ( in | out ) feature
  enclosing_feature_prototype_name

<port description>:
(in | out | in out) (event port | data port | event data port)
  <access description>:
(requires|provides)(bus|data|subprogram group|subprogram) access
```

Examples of prototype bindings are shown in Listing 12-12. This table contains component classifier extensions of the process type *basic* and process implementation *basic.impl* that were introduced in Listing 12-10. In the process type extension *PID_basic*, the process type *basic* is supplied with the following prototype actuals. The prototype *rd* is bound to the data type *set_point* and the prototype *cmd* is bound to the feature group *cmd_signals*.

In the process implementation extension *PID_basic.impl*, the process implementation *basic.impl* is supplied with the following prototype actuals. The prototype *cl* is bound to the thread implementation classifier *control.impl*, and the prototype rt is bound to the thread group implementation *control_law_grp.impl*. The feature prototype *cd*, which is a prototype of the process type *basic*, is bound to an in data port with the data classifier *speed_control_data*.

The system implementations shown in Listing 12-12 illustrate the binding of prototype actuals as part of subcomponent declarations. For the process subcomponent *PID_laws* of the system implementation *controller.impl* the binding of the prototypes of the process implementation classifier *basic.impl* are included as part of the classifier reference. For the prototype *rd* the prototype actual is declared to be the prototype *rd* of the system type *control*. In this case, the actual data classifier *set-point* is provided to the system subcomponent *control_subsystem* in the system implementation *top.impl*. This ensures that the same data type is used for the port *input* of the system *control_subsystem* and for the port *input* of the process *PID_laws*.

Listing 12-12: *Example Prototype Bindings*

```
process PID_basic extends basic
    (rd => data set_point, cmd => feature group cmd_signals)
end PID_basic;

process implementation PID_basic.impl extends basic.impl
  (cd => in data port speed_control_data,
   cl => thread control.impl,
   rt => thread group control_law_grp.impl )
end PID_basic.impl;

system controller
prototypes
  rd: data;
features
  input: in data port rd;
end controller;

system implementation controller.impl
subcomponents
    PID_laws: process basic.impl (rd => data rd,
      cmd => feature group cmd_signals,
      cd => in data port speed_control_data,
```

```
      cl => thread control.impl,
      rt => thread group control_law_grp.impl);
connections
  Inconn: port input -> PID_laws.input;
end controller.impl;

system top
end top;

system implementation top.impl
subcomponents
  control_subsystem: system controller.impl
    (rd => data set_point);
-- other subsystems and connections
end top.impl;
```

12.4.4 Properties

The *Prototype_Substitution_Rule* property specifies the rule that is used to determine what an acceptable prototype actual is. The rules are *Classifier_Match*, *Type_Extension*, and *Signature_Match*. This property can be associated with a prototype declaration or with the enclosing component type or component implementation to indicate that it can be inherited by all prototypes. The substitution rules for prototypes are summarized in Table 12-3. They correspond to the classifier substitution rules in Section 12.3.3.

Table 12-3: *Prototype Substitution Rules*

Value	Description
Classifier_Match	The classifier provided as prototype actual must be identical to the classifier referenced in the prototype declaration. If the declared prototype is abstract, any classifier can be used. If the prototype declaration specifies only a component type, any implementation of that type is acceptable. (This is the default value.)
Type_Extension	Any type extension of the component type declared with the prototype is an acceptable prototype actual.
Signature_Match	The classifier provided as prototype actual must match the signature of the classifier referenced by the prototype declaration. For a component type, the component type of the new classifier must have the same set of features, modes, and flows as specified in the prototype. For a component implementation, the component type must match and the implementation subclauses must match (e.g., subcomponents).

Chapter 13

Annotating Models

AADL allows you to annotate your model in several ways. You can document your model using comments, you can add information about a model element using properties, and you can annotate classifiers using annex sublanguages. We describe the use of comments and properties in this chapter and introduce annex annotations in the next chapter.

13.1 Documenting Model Elements

AADL offers several ways of documenting components of a system. In previous chapters, you have learned how to use features to specify the interface of components, and properties to record characteristics relevant to analyses. In this section, you will learn about comments, and about the ability to record that you intentionally left a section of a component specification empty.

13.1.1 Comments and Description Properties

You can include comments with any model element. This means you can add comments in either the graphical view or textual representation and they are visible in the other. In the graphical view, you do so by selecting the model element in a graphical editor and fill in the comment field for the model element. In the textual representation, you do so by prefacing the comment with two hyphens (--). All text on that line following these hyphens is treated as a comment. You can place a comment on the same line after a declaration. This is shown in Listing 13-1, where the comment "a thread instance." is contained on the same line as the thread subcomponent declaration. AADL does not have a special multiline comment symbol. This means that each line containing a comment must have the comment symbol, as shown in the first four lines in Listing 13-1. AADL does not place any restrictions on the formatting of comments.

Listing 13-1: *Example Comments in an AADL Model*

```
--
-- The process controller.speed provides the basic
-- speed control functions for the system.
--
process implementation controller.speed
subcomponents
control_laws: thread c_laws.speed; -- a thread instance
end controller.speed;
```

You can also define a *Description* property that has a string value and is applicable to all elements of an AADL model. You can then use it to provide additional documentation. You can further provide properties as to the authorship of the component and other relevant documentation.

13.1.2 Empty Component Sections

Component type, component implementation, feature group type, and package declarations consist of several sections that are optional. As you learned earlier, you can partially specify these model elements and later fill in missing sections during refinement. If you want to indicate that a part is intended to be empty, you can use a none statement (**none;**). This allows a consistency checker to determine whether you accidentally left a part empty. The declarations and the specific sections in which a **none** statement can be used are summarized in Table 13-1.

Table 13-1: *None Statement Options*

Declaration	Sections That Can Include a none Statement
Component type	Prototypes, features, flows, modes, properties
Component implementation	Prototypes, subcomponents, calls, connections, flows, modes, properties
Feature group	Prototypes, properties
Package	Properties

13.2 Using Properties

Properties define characteristics for the elements that comprise an AADL model. These elements include those declared using AADL standard constructs, such as processors, threads or ports, as well as elements declared as part of an annex such as error states and error events in the Error Model Annex. Each property has a name and a type. Its definition also specifies which model elements the property applies to. A property type establishes the values that can be assigned to a property.

There are AADL standard (predeclared) properties and property types. Collectively, these standard properties and property types encompass common attributes for the elements of the language. For example, a standard property of a port is *Required_Connection*, which is of standard type **aadlboolean** and has a value of true or false. Its default value is true. However, you can use a property association to assign the value false to this property for a port, allowing that port to be unconnected. The predeclared properties are organized into a collection of property sets and can be referenced without identifying the property set name.

The predeclared properties of the AADL standard are organized into the following property sets:

- *Thread_Properties* related to tasks and their interfaces
- *Timing_Properties* related to task execution and communication timing
- *Memory_Properties* related to the use of memory as storage and data access
- *Programming_Properties* related to mapping application components to source code
- *Deployment_Properties* related to the binding of application software to computer hardware
- *Modeling_Properties* related to the model itself

The AADL also permits users to define additional properties, property types, and property constants through additional property sets. The language constructs for doing so are discussed in Chapter 14. Property sets have been introduced as part of Annex standard documents. For example, the Error Model Annex standard [EAnnex] defines a property to specify the probability of occurrence of faults. Similarly, the Data Modeling Annex standard [DAnnex] defines properties for data components to characterize relevant aspects of data models (e.g., base types, type constructors such as record or array, or measurement units). The data model itself may exist in a data modeling language such as UML class diagrams, ASN.1, or in data types of programming languages. Other property sets are introduced with analysis tools in support of certain system analyses. For example, to support a security analysis plug-in to the OSATE toolset the Software Engineering Institute has introduced a collection of properties to specify security levels, categories of documents with different security requirements, and policies for handling declassification [Hansson 08].

13.2.1 Assigning Property Values

You assign values to properties and associate them with model elements through property associations. A basic property association consists of a property name, an assignment (association) operator, an optional **constant** keyword, and the value or list of values to be assigned to the property. The template for a basic property association declaration is shown in the following box.

```
<property name> => [constant] <property value>;
```

For predeclared properties you can just use the property name. For other properties you must include the name of the property set in which you defined the property (for example, *Property_Set_ Name::property*). The property value can be a single value whose type must match the type specified by the property definition, or it can be a list of values of the same type if the property has been defined to accept such a list. The list of values is a comma-separated list placed in parentheses.

With the **constant** keyword, you indicate that the property value assigned through this property association cannot be changed. For example, when you declare a property association of *Period* => **constant** *20 ms* for a thread type, it becomes the *Period* value for all instances of this thread type, and it cannot be changed by a thread implementation, subcomponent, or specific instance of this type. If the property association is declared without the **constant** keyword, you can assign a different value for different implementations of this thread type, or even different values for different instances. For more on how property values are determined, see Section 13.2.3.

A few basic example property associations are shown in Table 13-2.

Table 13-2: *Basic Property Association Examples*

Example	Description
Deadline =>50 ms;	Assigns the value 50ms to the predeclared standard property *Deadline*—no property set name is required
Error_Set:: Max_errors => 3;	Assigns the value 3 to the property *Max_Errors* that is defined in the property set *Error_Set*—the property set name is required
Source_Text =>("control_laws.java", "control_equations.java");	Assigns two source file names as the list of files containing source code
Error_Set::Error_kind => DivideByZero;	Assigns the value of a user-defined enumeration to the property *Error_kind*
Compute_Execution_Time => 2 ms..3 ms;	Assigns the range of 2ms to 3ms to the standard property *Compute_Execution_Time*

In the case of properties that accept lists of values you can specify a single value without placing it within parentheses. For such properties you can also append values to a previously assigned list of values. You do this with the operator +=>. For example, you may have specified a list of C header files for a process type with the *Source_Text* property. You can add C implementation files to each of the process implementations that contain the respective source code.

You can also assign property values that only hold under certain conditions. For example, you can assign different execution time values for a thread when it executes under different modes. You do this by constraining the value with an **in modes** statement as shown in Table 13-3.

You can also assign property values to software components that only hold when the component is involved in binding to a particular hardware component. For example, you can assign different execution times to a thread depending on whether it is bound to one type of processor or a different type of processor by using the **in binding** statement. An example is shown in Table 13-3.

Table 13-3: *More Property Association Examples*

Example	Description
Source_Text +=>"control_laws.c";	Assigns an additional source file name to the list of files containing source code
Compute_Execution_Time => 5 ms..10 ms in binding (PC.marine_certified);	Assigns the range of 5ms to 10ms to the standard property *Compute_Execution_Time* in cases when the thread is bound to an instance of the processor implementation *PC.marine_certified*
Compute_Execution_Time => *2 ms..3 ms in modes (initialization, shutdown),* *10 ms..20 ms in modes (nominal)*	Assigns the range of 2ms to 3ms to the standard property *Compute_Execution_Time* in the modes *initialization* and *shutdown*, and the range of 10ms to 20ms in the *nominal* mode

13.2.2 AADL Property Types and Values

Each property has a type that defines the specific values that can be assigned to the property. The various values that can be assigned to properties and example property associations for those types are summarized in Table 13-4.

Table 13-4: *A Summary of Property Values*

Value	Descriptions and Examples
Boolean (*aadlboolean*)	Boolean value (true or false) *Required_Connection => **true**;*
Real numeric (*aadlreal*)	Real numeric values with or without units The property definition specifies whether a unit is expected *Custom_Property_Set::boat_length => 3.0 meters;*
Integer (*aadlinteger*)	Integer values with or without units The property definition specifies whether a unit is expected *Period => 20 ms;*
String (*aadlstring*)	Alphanumeric string values *Source_Name => "ScaleData";*

Table 13-4: *A Summary of Property Values (continued)*

Value	Descriptions and Examples
Enumeration	A literal element from a set of literals defined by an enumeration type *Memory_Protocol => read_only;*
Unit	A measurement unit identifier from a set defined by a unit's type The unit's type includes conversion factors between the units *PBA_Property_Set::lengths_unit_info => feet;*
A range of real values	A range of real values with or without units *Physical::pressure_tolerances => 2.0 psi..3.0 psi delta 0.5;*
A range of integer values	A range of integers with or without units *Compute_Execution_Time => 1 ms..5 ms;*
Property value	The value of the named property is assigned (the property types must match) *Deadline => Period;*
Property constant	The value of the named property constant is assigned (the property types must match) The property constant is defined in a property set *Period => SystemParameters::BaseRate;*
Classifier	Values that are component types, component implementations, feature group types. The example property expects a list of classifier values. *Allowed_Processor_Binding_Class =>* *(**classifier**(marine_certified));*
Reference	Values are contained model elements (e.g., a subcomponent). The example property expects a list of reference values. *Allowed_Processor_Binding => (**reference**(compute_platform.* *RT_1GHz));*
Record	Values that have multiple property values (fields) within a single structure *ErrorAlarms => [AlarmCount => 5;* *AlarmType => transient;];*
Computed	A user-defined function that calculates a property value based on the given model element *physical::GrossMass => **compute** (calc_mass);*

The numeric values of integer and real or their ranges must be specified with a unit if the property definition states that the property requires a unit.

The range of integer or real values represents closed intervals of numbers that includes a lower bound of the interval, an upper bound of the interval, and (optionally) a delta indicating the difference between adjacent values. The delta may be unspecified as shown in Table 13-4.

Property constants provide a simple way of parameterizing the values assigned to properties in different parts of a model. Constants can be used anywhere a value of the same type is expected. For example, the property associations in Listing 13-2 assign values to the *Period* and *PBA::Max_Control_Dimensions* properties of the thread *algorithm.impl*. In the property set timing, the property *HiRate* is defined as a *constant* of the type *Time* with a value of 5ms. Time is a standard property type whose values are integers with time units. A change in the period of all instances of this thread type (as well as any other thread type with the same property association) can be accomplished simply by changing the value of *HiRate* in the property set. Similarly, you can parameterize the value of an integer property *PBA::Max_Control_Dimensions* with the integer constant *Max_Size*. Please note that property constants cannot be defined for reference value or classifier values.

Listing 13-2: *Property Associations That Reference a Property Constant*

```
thread implementation algorithm.impl
properties
Period => timing::HiRate;
PBA::Max_Control_Dimensions =>  PBA::Max_Size;
end algorithm.impl;


--

property set timing is
HiRate: constant Time => 5 ms;
end timing;

property set PBA is
Max_Control_Dimensions: aadlinteger applies to (all);
Max_Size constant aadlinteger => 5;
end PBA;
```

Property values that are reference values refer to specific instances of model elements using the containment hierarchy. You specify the path of the reference relative to the location at which the property association is declared. You have seen the use of reference values in software to hardware binding declarations such as *Actual_Processor_Binding*.

The property type *computed* allows you to specify that the value of a property is determined by invocation of a user defined function whose parameter is the model element for which the property values is retrieved. This function, available in a source code library to the tool environment, can for example calculate the value of the property for a component by summing up its subcomponent property values. In the example of Table 13-4 the function is assumed to add up the net mass of all subcomponents and the component itself to determine its gross mass.

13.2.3 Determining a Property Value

AADL has rules for determining the property value of a model element. For example, if you associate a *Period* property value of 20ms with a thread type, all instances of this type have this property value. In this section, you will learn what these rules are.

A property value for a model element is determined according to the following rules (in order):

- According to the highest contained property association in the hierarchy (see Section 13.2.4); if not present then
- According to basic property associations; if not present then
- By inheritance from enclosing components if the property is defined as **inherit**; if not present then
- The default value specified as part of the property definition; if not present then the value is considered to be undefined.

You can declare basic property associations in the properties section of component types, component implementations, feature group types, and packages. You can also declare basic property associations as part of declarations of subcomponents, features, and other elements inside component types and implementations, and feature group types by placing them in curly brackets.

The property value of an instance model element is determined by the property value of the subcomponent, connection, feature, flow specification, end-to-end flow, mode, or mode transition that has been instantiated. Table 13-5 defines the rules for determining these property values. If no property value is found according to this table, the inheritance and default value rules are used.

Table 13-5: *Summary of Property Determination Rules*

Rule	Find the Property Value
Subcomponents	In the subcomponent declaration; if not found In the subcomponent being refined, if declared with **refined to**; if not found In the component implementation or component type referenced by the subcomponent declaration
Features	In the feature declaration; if not found In the feature declaration being refined, if declared with **refined to**; if not found In the classifier referenced by the feature (such as the data component classifier for data and event data ports; or the feature group type for feature groups)
Feature groups	In the feature group declaration; if not found In the feature group declaration being refined, if declared with **refined to**; if not found In the feature group type referenced by the feature group
Connections	In the connection declaration; if not found In the connection declaration being refined, if declared with **refined to**
Flow specs	In the flow spec declaration; if not found In the flow spec declaration being refined, if declared with **refined to**
End-to-end flows	In the end-to-end flow declaration; if not found In the end-to-end flow declaration being refined, if declared with **refined to**
Modes	In the mode declaration; if not found In the mode declaration being refined, if declared with **refined to**
Mode transitions	In the mode transition declaration; if not found In the mode transition declaration being refined, if declared with **refined to**
Subprogram calls	In the subprogram call declaration; if not found In the classifier or subprogram feature referenced by the call
Subprogram call sequence	In the subprogram call sequence declaration
Component implementations	In the component implementation properties section; if not found In the component implementation being extended, if **extends**; if not found In the component type associated with the component implementation

Table 13-5: *Summary of Property Determination Rules (continued)*

Rule	Find the Property Value
Component types	In the component type properties section; if not found In the component type being extended, if **extends**
Feature group type	In the feature group type properties section; if not found In the feature group type being extended, if **extends**
Package	In the package declaration properties section

13.2.4 Contained Property Associations

Contained property associations provide a way of attaching property values to model elements within the hierarchy of a model instance. You do this by declaring the property association at a higher level of the model and indicate a path to the model element to which the property value belongs. The **applies to** statement allows you to specify multiple paths in a comma-separated list as shown in the following box:

```
<property name> => <property value> applies to <path>
    ( , <path> )*;
```

You can declare contained property associations at any level of the model hierarchy. If more than one contained property association attaches a property value to a model element, the contained property association declared at the highest level of the hierarchy takes precedence.

With contained property associations, you can define configuration parameters for a system at a single point (e.g., at the highest point possible in the component hierarchy). In that way, the parameters are in a centralized location for elements of a model and you can readily identify, adjust, and review them. This central location can be used even for elements that are deeply nested within a component. In Section 11.2.1 you used contained property associations when specifying deployment configurations as bindings of software components to hardware components.

Contained property associations are required when a property value involves a reference to another part of a model. In these associations, the property value that is assigned to the property is a reference to an element within the model. For example, the binding property of a thread must reference (identify) the processor to which it is bound. To accomplish this, the property association must be declared as contained property association attached to a model component that is the common parent of the component being referenced and the component to which the property value belongs.

13.2.5 Determining the Property Value: An Example

Listing 13-3 shows an example with contained property associations as well as basic property associations. Two contained property associations are declared within the system implementation *cc_complete.impl*. In the first association, the computation time for the compute entry point of the subcomponent thread *control_algorithm* is assigned the range of *2 ms..5 ms*. The thread *control_algorithm* is contained within the process *control_laws* that is a subcomponent of the system *cruise_control*. In the second association, the *Required_Connection* property is assigned the value *false* for the out data port *out_port* of the thread *adjust*, which is contained in the *control_laws* instance of process implementation *control.impl*. The process *control* has a contained property association for the *Required_Connection* property that applies to one of its own ports. The *Period* property is declared for the system *cruise_control* with the value 20ms. Since it is an inheritable property this value is the default for all threads that do not have their own property association. In the case of the *adjust* thread the *Period* property value is 10ms because the subcomponent declaration for this thread includes a property association. The graphical representation shown in the lower portion of that table highlights the assignment of property values relating to the thread subcomponents *adjust* and *control_algorithms*.

Listing 13-3: *Contained Property Associations*

```
system cc_complete
end cc_complete;
--
system implementation cc_complete.impl
subcomponents
brake_pedal: device brake_pedal;
cruise_control: system cruise_control.impl;
throttle_actuator: device throttle_actuator;
connections
C1: port brake_pedal.brake_event -> cruise_control.brake_event;
C2: port cruise_control.throttle_setting -> throttle_actuator.
throttle_setting;
properties
Compute_Execution_Time => 2 ms..5 ms applies to
    cruise_control.control_laws.control_algorithm;

Required_Connection => false applies to
    cruise_control.control_laws.adjust.out_port;
```

```
end cc_complete.impl;
--
system cruise_control
features
brake_event: in event data port;
throttle_setting: out data port;
properties
Period => 20ms;
end cruise_control;
--
system implementation cruise_control.impl
subcomponents
data_in: process interface;
control_laws: process control.impl;
connections
C1: port brake_event -> data_in.brake_event;
C3: port data_in.out_port -> control_laws.in_port;
C5: port control_laws.out_port -> throttle_setting;
end cruise_control.impl;
--
process control
features
in_port: in data port ;
out_port: out data port ;
properties
Required_Connection => false applies to out_port;
end control;
--
process implementation control.impl
subcomponents
adjust: thread adjust_sensor_value.impl {Period => 10 ms;};
control_algorithm: thread algorithm.impl;
end control.impl;
--
thread adjust_sensor_value
features
in_port: in data port;
out_port: out data port;
end adjust_sensor_value;
--
thread implementation adjust_sensor_value.impl
end adjust_sensor_value.impl;
--
thread algorithm
features
in_port: in data port;
out_port: out data port;
end algorithm;
--
thread implementation algorithm.impl
end algorithm.impl;
```

continues

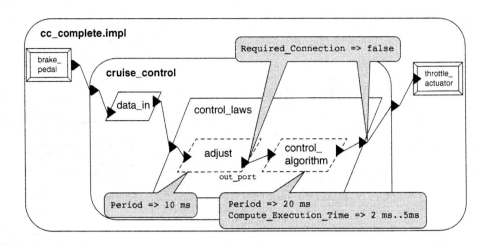

Chapter 14

Extending the Language

AADL allows you to introduce additional properties and property types through property sets. This is useful if you need to add information to an AADL model and no appropriate predeclared or user defined property has been declared. In addition, you can introduce a new sublanguage to AADL as an annex extension. You can then use this sublanguage to define reusable annex libraries similar to collections of classifiers, and attach sublanguage expressions to classifiers through annex subclauses. You will want to use sublanguages if you need to introduce objects that can have properties. For example, the Error Model Annex has been introduced as a sublanguage standard, such that you can define faults and error states as objects and attribute them to a component and associate property values with them (e.g., probability of occurrence).

14.1 Property Sets

You define new properties, property types, and property constants in property sets. Each property set provides a separate name space. Property set names are simple identifiers (i.e., there are no nested property set names). By preceding the property name, property type name, or property constant name with the property set name followed by "::" you can uniquely reference them. You have to include the property set name even if the property, property type, or property constant is declared in the same property set. For predeclared properties, property types, and constants you may omit the property set name.

If you refer to properties, property types, or property constants in another property set, you must declare the property set as accessible using a **with** declaration. Similarly, you must add a package name to the **with** declaration if property definitions reference a classifier in that package. Finally, you have to add the property set name to the **with** declaration of a package if the package contains references to properties, property types, or property constants of that property set. In other words, you explicitly declare your intent to reference entities in a package or a property set by adding the package name or property set name to the **with** declaration of the package or property set that contains the reference.

14.1.1 Declaring Property Sets

The template for declaring a property set is shown in Figure 14-1. The property set name must be a legal AADL identifier. You can declare a property set that is initially empty, to be filled in by somebody else or later by you. The optional **with** import statement identifies the other property sets and packages that are referenced from within the property set. The property type, property, and property constant declarations can be included in any order. In other words, there is no requirement to place the property type declaration before the property definition declaration that references it.

Table 14-1 summarizes the declarations allowed in a property set. Each is discussed in more detail in subsequent sections.

```
property set  <property set name> is
   [with <property set or package name(s)>]
<property type declaration>
<property declaration>
<property constant declaration>
end <property set name>;
```

Figure 14-1: *A Property Set Declaration Template*

Table 14-1: *Declarations Within a Property Set*

Declaration	Description
Property type	Defines a type for the values that are acceptable to a property. You define a property to be of a specific property type, thus limiting the values that this property accepts in property associations.
Property definition	Defines a property. You do this by declaring a name and a type for the property and identify which AADL model elements (e.g., mode, feature group, flow, port, component, or connection) accepts values for this property through property association.
Property constant	Defines a symbolic name for a property value. You can reference this name in property declarations wherever the value itself is permissible.

Listing 14-1 shows a property set declaration *set_of_faults* and includes examples of property name, property type, and property constant declarations. The property named *comm_error_status* is defined as a property of type **aadlboolean** (*true* or *false*) that applies to system and device components. A property type *Speed_Range* is defined as a range of real values from *0.0 mph..150.0 mph*. The property constant *Maximum_Faults* is defined as the integer value *3*.

Listing 14-1: *Sample Property Set Declarations*

```
system implementation data_processing.accelerometer_data
properties
     set_of_faults::comm_error_status => true;
end data_processing.accelerometer_data;

property set set_of_faults is
-- A property definition declaration.
comm_error_status: aadlboolean applies to (system, device);
-- A example property type declaration.
Speed_Range : type range of aadlreal 0.0 .. 150.0 units (mph);
-- An example property constant declaration.
Maximum_Faults : constant aadlinteger => 3;
end set_of_faults;
```

14.1.2 Property Type Declarations

AADL has a set of built-in types for properties. These are shown in Table 14-2 and can be used immediately in property definitions. In

Table 14-2: *Built-In AADL Property Types*

Type	Descriptions and Examples
Boolean (*aadlboolean*)	Two truth values: *true, false*
String (*aadlstring*)	A possibly empty sequence of characters between two quotation marks. Double quotation marks are used to represent a quotation mark in a string. An end-of-line cannot appear in a string (i.e., the opening and closing quotation mark must be on the same line). *"", " ", "A", """", "Message of the day:"*
Real numeric (*aadlreal*)	A real value with one digit before the decimal point and at least one digit after the decimal point. A single underscore is permitted between any two digits for readability. The value can have a positive or negative exponent. *0.0, 0.125, 75.0, 3.14159_26, 2.6E6, –5.7E–6*
Integer (*aadlinteger*)	An integer value of one or more digits. A single underscore is permitted between any two digits for readability. The value can have a positive exponent. The value can be in a base other than base 10—expressed by a base between 2 and 16 followed by the value enclosed in # using digits 0 through 9 and A through F. *0, 1E6, 123_456, 16#FF#*

addition, AADL allows you to define your own types, just as you do when defining data types in programming languages. For that purpose, you use property type declarations.

With a property type declaration, you define your own property type by establishing a name and the set of legal values for a property of that type through a type definition. The template for a property type declaration is shown in the following box:

```
name: type  <type definition>;
```

The entry <type definition> can be one of the built-in property types or it can be defined with one of the type constructors shown in Table 14-3. A type constructor can refer to another property by name. In this case, you must include the property set name separated by double colon (":::") to qualify the property type, even if the referenced property type is in the same property set. Only predeclared property types can be referenced without qualifying them by property set name.

Table 14-3: *AADL Property Type Constructors*

Type	Acceptable Values and Examples
Real numeric subset	Real values within a specified range *aadlreal –100.0 .. 100.0*
Integer subset	Integer values within a specified range *aadlinteger 1..12*
Unit	Identifiers from the defined list of measurement unit identifiers. Any AADL identifier is an acceptable unit identifier. Unit identifiers other than the first must have a conversion factor. The conversion must be in terms of a unit declared earlier in the list. *units (meters, km => meters*1000)* Example: *Metric_Length:* **type units** *(meters, km => meters*1000);*
Real numeric with units	Real values with measurement unit. *aadlreal units (feet)* Example: *Height:* **type aadlreal units** *<PSName>::Metric_Length* Note that the type constructor for height refers to the name of a units type, which must include the property set name.
Integer with units	Integer values with measurement unit *aadlinteger units (second, minute => second * 60)* *aadlinteger 0 .. 250 units (kph)*
Range of real values	Values that represent a closed interval of real numbers that includes numbers from the lower bound to the upper bound of the interval. It may include measurement units. The interval may be limited to be within the specified range of values. *range of aadlreal* *range of aadlreal 0.0..5.0* *range of aadlreal units (psi)* You can reference a numeric type in the range type constructor. This reference must be qualified by the name of the property set it is declared in. *range of <PSName>::Height* An example of an acceptable property value of the last range type is *0.0 meters .. 3.0 meters*

continues

Table 14-3: *AADL Property Type Constructors (continued)*

Type	Acceptable Values and Examples
Range of integer values	Values that represent a closed interval of integers that includes integers from the lower bound to the upper bound of a interval. It may include a measurement unit. The interval may be limited to be within the specified range of values. *range of aadlinteger 1..5;* *range of aadlinteger 0 .. +100 units (mpg);*
Enumeration	Identifiers from an ordered list of enumeration literals. Any legal AADL identifier is an acceptable literal. *enumeration (bad_value, no_value, OK)*
Classifier	Classifier references whose category matches one of the categories in the specified list. The category must map into one of the AADL Meta model classes that represent classifiers. For the constructor example shown, any thread type or thread implementation can be referenced, but the system classifier reference is limited to system types. *classifier (thread, thread group, process, data, system type)*
Reference	References to named model elements that are instances of one of the AADL meta model classes in the specified list or a subclass of these classes. For the constructor example shown, references can be to data components and to any kind of ports. *reference (data, port)*
Record	A collection of property values (fields), each accessible by name within a single structure. Any property definition is an acceptable definition of a record field. *record (name: aadlstring, initial: aadlreal, latest: list of aadlreal)*

In the examples shown in Listing 14-2, the property type *bit_error* is defined as an **aadlboolean**. The predefined **aadlboolean** property type has two legal values, true and false. The property types *fault_category* and *fault_condition* are defined as enumeration types. An enumeration property type defines a specific set of identifiers as its legal values.

Numeric type declarations can have measurement units and limits on acceptable values. For example, the type *number_of_components* is declared in the property set *some_types* as an **aadlinteger** that accepts values between 0 and 25. The property type *metric_length_units* introduces metric measurement units for length with conversion factor. Within the property set *marine_measures*, the *Depth_Units* units type is defined. It is used to define the type *working_depth* as a range of real values with any of the *Depth_Units* units. You have to qualify the reference to *Depth_Units* with the property set name. The property type *Metric_Length* is defined as real values with units defined in the property set *some_types*. Therefore, you have to include that property set in the *with* clause of *marine_measures*.

Listing 14-2: *Sample Property Type Declarations*

```
property set set_of_faults is
bit_error: type aadlboolean;
fault_category: type enumeration
      (benign, tolerated, catastrophic);
fault_condition: type enumeration (okay, error, failed);
end set_of_faults;

property set some_types is
number_of_components: type aadlinteger 0 .. 25;
metric_length_units: type units
      (meters, kilometers => meters * 1000);
end some_types;

property set marine_measures is
with some_types;
  Depth_Units : type units ( feet, fathom => feet * 6,
                             shackle => fathom * 15 );
  working_depth: type range of aadlreal
      marine_measures::Depth_Units;
  Metric_Length: type aadlreal some_types::metric_length_units;
  end marine_measures;
```

14.1.3 Property Definitions

A property definition declaration allows you to define a new property by declaring a name, and by specifying a property type. The property is specified either by a type constructor or by reference to a property type by name. You indicate acceptable owners of the property through the **applies to** clause. You can declare the property to accept a single value or a list, whose values can be single values of the specified type or can be lists of values themselves. The templates for both single-valued and list-valued property definitions are shown in the following box.

```
name : [inherit] <property type> [=> <default value>]
      applies to <property owner>;
name : [inherit] ( list of )+ <property type>
      [=> <default values>] applies to <property owner>;
```

You can limit the property ownership to certain model elements (i.e., to components of specific categories by naming their AADL Meta model class), to components of specific classifiers, and to other named model elements through their AADL Meta model class, such as ports, connections, modes, or even model elements defined in Annex sublanguages (see Section 14.2). Alternatively, you can use the reserved word **all** to indicate that the property is acceptable to all named model elements. Note that all properties, both standard and newly defined, apply to **abstract** components, regardless of the **applies to** statement used in their definition.

The optional reserved word **inherit** indicates that the property can inherit a value from a component in which it is contained.

The optional default value is returned as property value if no value is explicitly assigned (see Section 13.2.3). If no default value is specified the value is undefined or the empty set in the case of a list of values.

Example property declarations within a property set *set_of_property_names* are shown in Listing 14-3. The property *critical_unit* is declared as type **aadlboolean** and applies to all component categories. Its property value can be inherited from the enclosing component if not assigned to a model element. The property *Boat_Length* is declared in terms of the property type *Metric_Length* defined in another property set. The property *Maintainers* accepts a list of string values, with a list of three values as default. The property value if a single string is assigned for the system *basic_dual*. The list-valued property *Shared_Memory* of type **reference**, which accepts references to memory components, is assigned a set of values in the system implementation *basic_dual.impl*.

Listing 14-3: *Sample Property Definition Declarations*

```
property set my_props is
with marine_measures;
critical_unit: inherit aadlboolean applies to (all);
Boat_Length: marine_measures::Metric_Length
    applies to (device, system);
Maintainers: list of aadlstring => ("Peter", "John", "Dave")
    applies to (all);
Shared_Memory: list of reference (memory) applies to (system);
end my_props;

package marine_lib
with my_props
system basic_dual
properties
myprops::Maintainers => ("John");
```

```
end basic_dual;
system implementation basic_dual.impl
subcomponents
mem01: memory;
mem02: memory;
mem03: memory;
properties
myprops::Shared_Memory => (reference(mem01), reference(mem03));
end basic_dual.impl;
device MotorBoat
properties
Boat_Length => 10.0 meters;
end MotorBoat;
end marine_lib;
```

14.1.4 Property Constant Declarations

Property constants are property values that are known by a symbolic name. You declare property constants in property sets. The name can be used wherever the value itself is permissible. Some property constants are provided in AADL standard property sets.

The basic declaration templates for single-valued and list-valued property constant declarations are shown in the following box. In the case of a list-valued constant the elements of the list can be single values or list values themselves.

```
name: constant <property type> => <property value>
name: constant ( list of )+ <property type> => <property values>
```

The <property type> is a type constructor or refers to an explicitly declared property type. If the reserved words **list of** are used in the declaration, the type is list valued.

Some sample declarations are shown in Listing 14-4, where, for the property set *limits_set*, *Max_Threads* is defined as an integer value of *256*; *Minimum_value* is defined as a real value of *5.0*; and *Default_Fault_State* is defined as a constant of the type *fault_condition* with the value of *okay*. The type *fault_condition*, referenced in Listing 14-4, is defined in the *package set_of_faults*, as shown in Listing 14-2.

Listing 14-4: *Sample Property Constant Declarations*

```
property set limits_set is
with set_of_faults, operations_set;
Max_Threads : constant aadlinteger => 256;
Minimum_value: constant aadlreal => 5.0;
```

continues

```
Default_Fault_State: constant
    set_of_faults::fault_condition => okay;
end limits_set;
```

14.2 Annex Sublanguages

Additions to introducing new properties into AADL, you can extend the set of language concepts by introducing sublanguages through Annexes. An Annex sublanguage can provide new categories of model elements, allow you to declare classifiers of those new model element categories, define properties that can be associated with them, and attach instances of the model elements to components. The structure of this sublanguage is specified in an annex document by defining a Meta model, whose classes are subclasses of the AADL Meta model, in particular of the classes *Named Element* and *Classifier*. The annex document also defines a textual syntax to be used in annex clauses of the textual AADL model. You use annex library statements to declare classifiers of annex categories in packages, the same way you declare component classifiers in packages. You can then reference these annex classifiers in annex subclauses that are attached to component types and implementations—similar to subcomponent declarations referencing component classifiers. Annex libraries and annex subclauses may contain references into the core AADL model elements.

The Error Model Annex standard [EAnnex] is an example of such a sublanguage. You can introduce error state machines as classifiers in annex library declarations of the Error Model. These state machines consist of error states and error transitions (similar to modes and mode transitions), error events that represent faults (similar to subcomponents), and error propagations that represent component errors that may affect components this component interacts with. The error events and error propagation can have an occurrence property to indicate the probability of a fault occurring or being propagated. You then associate an instance of this error state machine with a component by referencing the error state machine classifier in an Error Model annex subclause. By doing so, you indicate that any instance of the component type or component implementation with this error state machine has a fault behavior described by the error state machine. In the Error Model annex subclause you can specify conditions under which errors are propagated out through a specific component port. This is an example, where an expression in an annex sublanguage can refer to a model element in core AADL.

AADL can have multiple Annex sublanguages. For example, the AADL standards committee is also introducing a Behavior Annex as a standard [BAnnex]. In addition, you can introduce nonstandard sublanguages that you may use only within your project.

14.2.1 Declaring Annex Concepts in Libraries

AADL has defined a standard syntactic construct through which you introduce annex libraries. The template for an annex library declaration is shown in the following box.

```
annex <annex name>  {** <library content> **};
```

Annex library declarations must be placed in packages. You can place them anywhere within the public or private section of the package. You can only declare one annex library for each sublanguage in each package. The annex name identifies the annex sublanguage and must be a legal AADL identifier. The library content is enclosed in the symbols {** and **}. It consists of the language constructs, such as classifiers that can be referenced by multiple annex subclauses. The annex sublanguage syntax can use anything other than the closing **}.

An excerpt from an SAE standard error model annex [EAnnex] library is shown in Listing 14-5. The library is declared in the package *E_Models* and identifies the sublanguage as *Error_Model*. *Error_Model* is the annex sublanguage name defined in the Error Model Annex standard. The excerpt establishes an error state machine model type *Three_State*. This error model type can be referenced in an error model annex subclause. Notice that the constructs for the error model annex are similar to the AADL language syntax and style. However, this need not be the case. Any set of textual language constructs can be included within an annex (e.g., the Object Constraint Language (OCL) or a temporal logic notation).

Listing 14-5: *An Example Error Annex Library*

```
package E_Models

public
annex Error_Model {**
error model Three_State
```

continues

```
features
Fail_Stop : error event
            { Occurrence => poisson 10E-4 } ;
Fail_Active : error event
            { Occurrence => poisson 10E-5 } ;
No_Data, Bad_Data : in out error propagation;
Error_Free: initial error state;
Stopped, Active_Fault: error state;
end Three_State;
                **};


end E_Models;
```

14.2.2 Using Annex Concepts in Subclauses

Annex sublanguage concepts introduced through an annex library are attached to an AADL model by using an annex subclause within a component type, component implementation, or feature group type declaration. An annex subclause must be the last section of either a component type or component implementation declaration following the properties section. A different annex subclause can be declared for each mode. A component type or component implementation may contain at most one annex subclause (per mode) for each annex sublanguage. The basic template for an annex subclause is shown in the following.

```
annex <annex name> {** < subclause content> **} [ <in modes> ] ;
```

The name of the annex name in the subclause identifies the annex sublanguage. The syntax of the subclause content can refer to items in the annex library and to basic AADL model elements.

An example annex subclause for the error model annex is shown in Listing 14-6. In this example, the error model type *Three_State*, defined in the annex library in package *E_Models* shown in Listing 14-5, is attached as the error model for the system implementation *basic.control*. In addition, values for the property *Occurrence* are assigned to error events within the error model.

Listing 14-6: *Example Annex Subclause Declarations*

```
system implementation basic.control
  subcomponents
    A: system sensors.speed;
    B: system controller.speed;
```

```
   annex Error_Model {**
      model => E_Models::Three_State;
      Occurrence => poisson 10E-3 applies to Fail_Stop;
      Occurrence => fixed 10E-4 applies to Fail_Active;
**};

end basic.control;
```

Chapter 15

Creating and Validating Models

Tools[1] are essential to the creation and validation of embedded system models. The AADL standard facilitates the integration of tools into a cohesive engineering environment through a standardized Meta model and XML interchange format specification for AADL.

15.1 Model Creation

You can create AADL models textually or graphically by using any of the AADL tools mentioned in Section 15.2. Figure 15-1 shows the layout of the user interface for OSATE. On the left you can see the workspace in a navigator view. It shows an AADL model organized as a single Eclipse project and split into a number of AADL packages. The middle of the figure shows both a graphical and a textual view of one of the packages. You can see a set of communicating threads as the implementation of the *flight manager* process. Across the top are a number of toolbars and a menu system for manipulating the AADL models, compiling them to check for semantic consistency, and for invoking a variety of analysis tools.

---

1. See the AADL Tools section of the AADL Public Wiki (https://wiki.sei.cmu.edu/aadl) for publicly available tools, including those mentioned in this section.

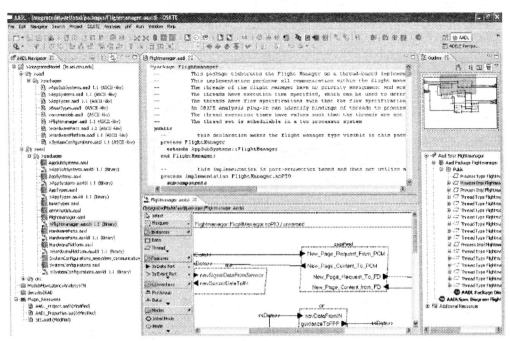

Figure 15-1: *Representation of an OSATE User Interface*

You can get started with creating AADL models in a number of different ways throughout the development life cycle. For example, as part of an architecture design review, you may convert drawings in a design document or PowerPoint slides into a high-level AADL model that you can then evolve over time. Just the creation of such a model in a strongly typed language with well-defined semantics allows you to discover potential inconsistencies in the model.

You may annotate the model with basic information from resource budget and capacity tables in a design document and perform a first level of resource budget analysis in the context of the model instead of a separate spreadsheet. You can do so in several ways. You can use the graphical or textual AADL model editor to find the threads to be annotated and fill in the property values through the AADL property view. You can add a forms-based front-end to the tool environment that simplifies this task and allows you to fill in a table prepopulated with threads to be annotated. You can also place all the timing properties in a single place as a collection of contained property associations similar to the binding configurations illustrated in Section 11.2.

You may be a member of a team to evaluate an existing system architecture. Again, you can capture information presented to you during an assessment in an AADL model and it can be a starting point for evolving it into a higher-fidelity model. If you have an existing implementation of a system, you may be able to extract architectural information from source code or design databases to populate an initial architecture model with detail and check for model consistency.

You can use AADL to represent a whole family of similar systems as a set of models that are refined from a common core [Feiler 07A]. Similarly, you can represent the reference architecture of a system in AADL, validate it, and then refine it into instantiations for specific applications. This has been done for a Mission Data System reference architecture from the Jet Propulsion Laboratory [NASA]. In these cases, you will make heavy use of component templates and refinement (Sections 12.3 and 12.4) as well as configuration of the model via properties through contained property associations (Section 13.2).

Other projects have used AADL to support the interaction between a system integrator and their suppliers. AADL models of subsystems are version controlled and interchanged through a model repository with the integrator validating the system architecture by integrating the subsystem models within AADL and validating them throughout the development life cycle at different levels of fidelity [SAVI]. In this case, you make use of separate public and private sections of packages to represent subsystem interfaces (Section 12.2), data types in separate packages playing the role of data dictionaries, and feature group types in separate packages representing the interface specification of data to be exchanged between subsystems (Section 10.4.1).

15.2 Model Creation Tools

In this section we summarize a number of tools that support creation of AADL models.

The Software Engineering Institute at Carnegie Mellon University developed the extensible Open Source AADL Tool Environment (OSATE). OSATE consists of a set of plug-ins for the open source Eclipse platform [Eclipse]. OSATE includes a compiler for textual AADL, a graphical editor for AADL, an instance model generator, and support for the XML-based XMI interchange format [W3C 04] for AADL based on its Meta model specification. The graphical Adele editor for OSATE has been provided by Ellidiss (www.ellidiss.com). A number of analysis and generation tools have been integrated with OSATE via the XMI interchange format (see Section 15.3). OSATE is integrated into the TOPCASED embedded system tool suite [TOPCASED].

STOOD is a commercial tool environment for embedded system development (www.ellidiss.com) that covers the complete life cycle from requirements to system build [STOOD]. Originally developed to support HOOD-RT, it has been extended to support AADL models. It provides a graphical frontend for architectural design and generates textual AADL. It includes an AADL Inspector for AADL model validation and a capability for extracting architecture information from source code. STOOD has been integrated with analysis tools and simulators as interfaced with other AADL tool chains including OSATE and Ocarina to take advantage of their analysis and generation capabilities.

The Rockwell Collins META toolset extends OSATE and includes a graphical editor for AADL based on a SysML editor, an architectural design pattern tool, a static model verification tool, and a compositional verification tool. The graphical editor has been interfaced with OSATE to convert between textual AADL models edited within OSATE and graphical models edited within the Enterprise Architect based SysML editor.

The ASSERT Set of Tools for Engineering (TASTE) toolset was developed by the European Space Agency (ESA) and its partners [TASTE]. It provides a set of graphical editors for creating the software architecture, the hardware architecture, and the deployment of the software on the hardware. These editors provide a simplified user interface to engineers, eliminating the need to learn the textual syntax of AADL.

UML tools using a UML profile for AADL can also create AADL models. A prototype implementation of such a profile was done for Rhapsody in the Aerospace industry initiative called System Architecture Virtual Integration (SAVI) [SAVI].

The AADL Inspector (www.ellidiss.com) is a standalone tool for checking the consistency of textual AADL model. It is integrated with the Cheddar scheduling analysis tool and a multiagent simulation engine for dynamic simulation of AADL models.

Organizations have created tools that generate AADL models from the content of design databases, and extract AADL models from existing Simulink models [AADLSimulink].

15.3 System Validation and Generation

Validation of a system architecture represented by an AADL model takes on several forms: validating the consistency of an AADL model, analyzing an AADL model with respect to functional and nonfunctional properties of the embedded system represented by the model, and generation of an implementation from such validated models.

AADL is a strongly typed language with a number of consistency rules. When you compile the textual representation of AADL or save changes to the graphical model of AADL, a semantic checker identifies inconsistencies in the model. Examples of inconsistencies include mismatches in the data type of connected ports, mismatches in the type of bus on bus access connections, and identification of multiple incoming connections on data ports—a race condition that results in inconsistent sampling of data from multiple sources.

The AADL standard also defines a set of consistency rules that are checked by consistency analysis tools. Examples of consistency checks include ensuring connections for ports that require connections, ensuring that port connections between threads on different processors are supported by a physical connection via buses, and that complete system models include both processors and memory to support threads and data components.

The real value of AADL comes from its well-defined semantics and the ability to annotate the architecture model with information relevant to different functional and nonfunctional properties, such as behavior, performance, timing, safety, reliability, and security. The annotated AADL acts as the reference source for different analytical models that are generated from the AADL model, such as timing models for scheduling analysis, cooperating state machine or Petri net models for model checking of behavior, fault trees, and stochastic models for safety and reliability analysis.

The Aerospace industry has recognized the need to maintain consistency between models and has put in place an Eclipse-based open source tool infrastructure [TOPCASED]. In addition, the aerospace industry is engaged in a multiyear initiative called System Architecture Virtual Integration (SAVI) to put into practice model-based architecture-centric engineering. This initiative is using AADL as a key technology for virtual system integration and analysis. This approach leads to discovery of system-level problems early in the life cycle and a major reduction in rework cost [SAVI]. A proof of concept demonstration project has shown the practicality of supporting modeling, validation, and generation of embedded software systems. This project has shown the ability to perform multitier modeling and multifidelity analysis of different system properties from the same architecture model, to perform incremental refinement and revalidation of both the embedded software system and the Mechatroncis aspects of an aircraft, and to support supplier and system integrator interaction in a distributed development setting.

15.4 System Validation and Generation Tools

The OSATE toolset includes a number of sample analysis tools to demonstrate analysis of multiple quality dimensions from the same architecture model. The Binpacker tool makes binding decisions for threads and data components taking into account binding constraints that you may have expressed through binding properties (see Section 11.2) and cluster threads such that network traffic is minimized [DeNiz 06]. Resource budget analysis is supported through a set of resource-related properties resource capacities and budgets for processor cycles, memory, and network bandwidth [DeNiz 08]. Early in the development process, you can assign resource capacities to hardware components and resource budgets to application components that use these resources. As the system model is refined by increasing the component hierarchy, you recursively divide up the resource budgets. At the task level, thread periods and execution times (the same data used for scheduling analysis), output rates and data sizes on port are used to compare the actual demand of software components against the budgets, and budgets totals are compared up the hierarchy and against capacities.

In a similar fashion, properties have been added to OSATE that characterize the resources of a physical system, such as mass and electrical power. Again, you assign property values to physical components represented by devices, buses, memory, processors, and their composites in the form of systems. You can then use an OSATE plug-in to perform these resource analyses, you can develop analysis scripts in a scripting language that has been integrated into OSATE, or you can export the data from OSATE in a spreadsheet format and process the data with commercial spreadsheet tools [SAVI].

If you are interested in integrating AADL models and Simulink design models, you can look at the SAVI demonstrations [SAVI]. In

SAVI, the Simulink models of physical system behavior in Mechatronic subsystems have been integrated with finite element models of physical structures such as wings via the AADL as common underlying architectural structure and interchange representation. In a different demonstration project you see the ability to extract AADL models from existing Simulink models, and the generation of an executable application code and runtime system from these validated models via Ocarina [AADLSimulink].

If you are interested in dependability, you can learn about how to add fault and fault propagation information into an AADL model through the Error Model Annex [Feiler 07]. You can introduce intrinsic types of errors, error state machines, and error propagations and associate them with component specifications and perform safety and reliability analyses, such as Functional Hazard Assessment (FHA), Failure Mode and Effect Analysis (FMEA), and Mean Time To Failure (MTTF) analysis, as has been demonstrated in [SAVI].

Similarly, you can deploy the Behavior Annex as a way of adding behavioral specifications to components that can be the basis for concurrency analysis and other forms of model checking. You can interface with various model checking tools via a common interchange format for model checkers called FIACRE, which can be generated from AADL models with behavior specifications [FIACRE].

You can use the Behavioral Language for Embedded Systems with Software (BLESS), which is an extension to the Behavior Annex, to specify assertions, including time-related assertions, about your system behavior and verify them with a theorem prover [BLESS].

The Application Specific I/O Integration Support Tool for Real-Time Bus Architecture Designs (ASIIST) [ASIIST] allows you to investigate potential resource contention issues in the computer platform. It performs schedulability analysis, bus delay analysis based on hardware flows, analysis of I/O cache interference effects, and comparative analysis across different model configurations.

If you are interested in code generation from AADL models combined with application component models expressed in C, Ada, Simulink, or Scade for distributed applications on several hardware platforms including operating systems with partition support, you may want to check out Ocarina [Ocarina]. Ocarina supports generation of runtime systems based on AADL models for single processors, networked processors, and ARINC653 based runtime architectures.

Ocarina is part of the TASTE toolset developed by the European Space Agency (ESA) and its partners [TASTE]. It allows software developers to integrate heterogeneous pieces of code produced either manually or automatically. Consistency of the integration is ensured through a data model in the ASN.1 language and behavior models expressed in SCADE and Simulink. TASTE also interfaces with the scheduling analysis tool MAST as well as a simulation tool.

ESA sponsored the COMPASS (Correctness, Modeling and Performance of Aerospace Systems) project [COMPASS], which created a tool chain supporting system-software co-engineering coherent set of specification and analysis techniques for evaluation of system-level correctness, safety, dependability, and performability of on-board computer-based aerospace systems. It utilizes the Error Model Annex of AADL to capture fault behavior and verify that faults are correctly handled by the fault management architecture.

The Rockwell Collins META toolset includes an architectural design pattern tool, a static model verification tool, and a compositional verification tool. It has been used to demonstrate the validation of systems based on compositional reasoning based on contract specifications [RC META].

If your concerns are security related, you can leverage prototype security plug-ins for AADL [Hansson 08]. One plug-in offers a high-level security analysis based on you assigning security levels to components throughout the component hierarchy. A second security analysis plug-in support detailed security analysis based on security models known as Bell LaPadula and Chinese Wall. In that case, you assign properties to components and to ports to provide security information regarding subjects, roles, and objects in terms of security levels and security categories.

In summary, many research and advanced technology groups have used AADL as a platform to interface their formal analyses capabilities with practitioners without exposing them to the intricacies of some of their formalisms. A current listing of extensions to OSATE as well as commercial and other open source AADL tools can be found on the AADL Web site (www.aadl.info) and on the public AADL Wiki (wiki.sei.cmu.edu/aadl). You can find presentations and tutorials for the various toolsets in the *User Days* section of the public AADL Wiki.

Appendixes

Appendix A

Syntax and Property Summary

A.1 AADL Syntax

The AADL syntax and grammar rules summarized in Listing A-1 conform to the AADL standard document [AS5506A]. The following table shows what some of the notations in the listing mean.

Summary of the Notation	
Bolded are AADL keywords	{ }* Repeatable
[ ] Optional term	{ }+ Repeatable once or more
| Alternative	::= Expansion term

Listing A-1: *Summary of AADL Language*

```
-- Packages to structure component specifications into libraries
AADL_model ::= package_spec | property_set

package_spec ::=
package package_name
( public package_content [ private package_content ]
```

continues

```
    | private package_content )
  [ properties ( { property_association }+ | none_statement ) ]
  end package_name ;

  package_name ::= { package_identifier :: }* package_identifier

  none_statement :: = none ;

  package_content ::=
  { name_visibility }* { classifier_declaration | annex_library }*

  classifier_declaration ::=
  component_type | component_implementation | feature_group_type

  name_visibility ::= import | alias

  import ::= with ( package_name | property_set_name )
      { , ( package_name | property_set_name ) }* ;

  alias ::=
   ( package_alias_identifier renames package package_name ; ) |
   ( [ component_type_alias_identifier renames
       component_category component_type_reference ; ) |
   ( [ feature_group_type_alias_identifier renames
       feature group feature_group_type_reference ; ) |
   ( renames package_name :: all ; )

  component_category ::=
  data | subprogram | thread | thread group | process | system |
  abstract | memory | processor | bus | device |
  virtual processor | virtual bus

  -- Component interface specification
  -- without extends it cannot contain refinements
  component_type ::=
  component_category component_type_identifier
  [ extends component_type_reference [ prototype_bindings ] ]
  [ prototypes ( { prototype }+ | none_statement ) ]
  [ features ( { feature }+ | none_statement ) ]
  [ flows ( { flow_spec }+ | none_statement ) ]
  [ ( modes ( { mode }+ { mode_transition }* | none_statement )
     | ( requires modes ( { mode }+ | none_statement ) ]
   [ properties
     ( { component_type_property_association }+ | none_statement ) ]
  { annex_subclause }*
  end component_type_identifier ;

  -- Component implementation blue print
  -- without extends it cannot contain refinements
  component_implementation ::=
  component_category implementation
      component_type_identifier.component_implementation_identifier
  [ component_type_prototype_bindings ]
```

```
[ extends component_implementation_reference
    [ prototype_bindings ] ]
[ prototypes ( { prototype }+ | none_statement ) ]
[ subcomponents ( { subcomponent }+ | none_statement ) ]
[ calls ( { subprogram_call_sequence }+ | none_statement ) ]
[ connections ( { connection }+ | none_statement ) ]
[ flows  ( { flow_implementation | end_to_end_flow }+
  | none_statement ) ]
[ modes ( { mode }+ { mode_transition }* | none_statement ) ]
properties
( { property_association | contained_property_association }+ |
    none_statement ) ]
{ annex_subclause }*
end component_type_identifier.component_implementation_identifier ;

-- Component classifier references
component_classifier_reference ::=
component_type_reference | component_implementation_reference

component_type_reference ::=
 [ package_name :: ] component_type_identifier

component_implementation_reference ::=
[ package_name :: ]
component_type_identifier.component_implementation_identifier

-- subcomponent instance in a component implementation
subcomponent ::=
subcomponent_identifier : [ refined to ] component_category
[ component_classifier_reference [ prototype_bindings ]
| prototype_identifier ]
[ array_dimensions [ array_element_implementation_list ] ]
[ { { subcomponent_property_association |
      contained_property_association }+ } ]
[ component_in_modes ] ;

array_dimensions ::=
    { array_dimension }+

array_dimension ::=
    [ [numeral | property_constant_identifier |
      property_identifier ] ]

array_element_implementation_list ::=
    ( component_implementation_reference [ prototype_bindings ]
      { , component_implementation_reference
        [ prototype_bindings ]  }* )

-- array selection used in contained property associations
array_selection_identifier ::=
    identifier { [ numeral [ .. numeral ] ] }+
```

continues

```
-- Component interaction points
feature ::=
abstract_feature | port | data_access | bus_access |
 subprogram_access  | subprogram_group_access |
 parameter | feature_group

-- Abstract feature can be refined into port,parameter, access,
-- and feature group features
-- Done by port/parameter/access/feature group refinement
-- referring to the abstract feature
abstract_feature ::=
abstract_feature_identifier : [ refined to ]
    ( [ in | out ] feature [ feature_prototype_identifier ] )
    [ array_dimension ]
[ { { feature_property_association }+ } ] ;

-- Interaction point for directional flow of data, events, and messages
port ::=
port_identifier : [ refined to ] ( in | out | in out )
( data port [ data_classifier_reference |
              data_prototype_identifier ]
| event port
| event data port [ data_classifier_reference |
                    data_prototype_identifier ] )
[ array_dimension ]
 [ { { port_property_association }+ } ] ;

-- Shared data access interaction point
data_access ::=
data_access_identifier :  [ refined to ]
 ( provides |  requires ) data access
      [ data_component_classifier_reference |
        data_prototype_identifier ]
[ array_dimension ]
 [ { { data_access_property_association }+ } ] ;

-- Bus access interaction point
bus_access ::=
bus_access_identifier :  [ refined to ]
  ( provides |  requires ) bus access
      [ bus_component_classifier_reference |
        bus_prototype_identifier ]
[ array_dimension ]
[ { { bus_access_property_association }+ } ] ;

-- Callable subprogram interaction point
subprogram_access ::=
subprogram_access_identifier :  [ refined to ]
    ( provides |  requires ) subprogram access
       [ subprogram_component_classifier_reference |
         subprogram_prototype_identifier ]
[ array_dimension ]
[ { { subprogram_access_property_association }+ } ] ;
```

```
-- Subprogram library access interaction point
subprogram_group_access ::=
data_access_identifier :   [ refined to ]
    ( provides |  requires ) subprogram group access
        [ subprogram_group_component_classifier_reference |
          subprogram_group_prototype_identifier ]
[ array_dimension ]
[ { { subprogram_group_access_property_association }+ } ] ;

-- Subprogram parameter
parameter ::=
parameter_identifier : [ refined to ]
 ( in | out | in out ) parameter
[ data_classifier_reference | data_prototype_identifier]
[ { { parameter_property_association }+ } ] ;

-- collection of features as one interaction point
feature_group ::=
feature_group_identifier : [ refined to ]
[ in | out ] feature group
[ [ inverse of ] feature_group_type_reference |
                 feature_group_prototype_identifier ]
[ array_dimension ]
[ { { feature_group_property_association }+ } ] ;

feature_group_type_reference ::=
[ package_name :: ] feature_group_type_identifier

-- type specification of a collection of features
feature_group_type ::=
feature group feature_group_type_identifier
[ extends feature_group_type_reference ]
[ prototype ( { prototype }+ | none  ) ]
[ features { feature }+ ]
[ inverse of feature_group_type_reference ]
 [ properties
( { feature_group_property_association }+ | none  ) ]
{ annex_subclause }*
end feature_group_type _identifier ;


-- Connections between features
connection ::=
abstract_feature_connection | port_connection |
parameter_connection | access_connection |
feature_group_connection

abstract_feature_connection ::=
abstract_feature_connection_identifier :
( feature source_feature_connection_reference (-> | <-> )
    destination_feature_connection_reference
| refined to feature )
```

continues

```
[ { { abstract_feature_connection_property_association }+ } ]
[ in_modes_and_transitions ] ;

feature_connection_reference ::=
    -- feature in the component type
    component_type_feature_identifier |
    -- feature in a feature group of the component type
    component_type_feature_group_identifier.feature_identifier |
    -- feature in a subcomponent
    subcomponent_identifier . feature_identifier |
    -- feature in a subprogram call
    Subprogram_call_identifier . feature_identifier

port_connection ::=
port_connection_identifier :
(( port source_port_reference (-> | <-> )
  destination_port_connection_reference )
| refined to port )
[ { { port_connection_property_association }+ } ]
[ in_modes_and_transitions ] ;

port_connection_reference ::=
    -- port in the component type
    port_identifier |
    -- port in a subcomponent
    subcomponent_identifier.port_identifier |
    -- port element in a feature group of the component type
    feature_group_identifier.element_port_identifier |
    -- data element in aggregate data port
    port_identifier. data_subcomponent_identifier |
    -- requires data access in the component type
    requires_data_access_identifier |
    -- data subcomponent
    data_subcomponent_identifier |
    -- data component provided by a subcomponent
    subcomponent_identifier.provides_data_access_identifier |
    -- data access element in a feature group of a component type
    feature_group_identifier.element_data_access_identifier |
    -- access to element in a data subcomponent
    data_subcomponent_identifier.data_subcomponent_identifier |
    -- processor port
    processor . processor_port_identifier |
    -- component itself as event or event data source
    self . event_or_event_data_source_identifier

parameter_connection ::=
parameter_connection_identifier :
( parameter source_parameter_connection_reference ->
    destination_parameter_connection_reference
| refined to parameter )
[ { { paramter_connection_property_association }+ } ]
[ in_modes_and_transitions ] ;
```

```
parameter_connection_reference ::=
  -- parameter in the thread or subprogram type or
  -- an element of that parameter's data
  parameter_identifier [ . data_subcomponent_identifier ] |
  -- parameter in another subprogram call
  subprogram_call_identifier.parameter_identifier |
  -- data or event data port in the thread type or
  -- an element of that port's data
  port_identifier [ . data_subcomponent_identifier ] |
  -- data subcomponent in the thread or subprogram
  data_subcomponent_identifier |
  -- requires data access in the thread or subprogram type
  requires_data_access_identifier |
  -- data access element in a feature group of the component type
  feature_group_identifier.element_data_access_identifier |
  -- port or parameter in a feature group of the component type
  feature_group_identifier [ . port_or_parameter_identifier ]

data_access_connection ::=
data_access_connection_identifier :
( data access source_data_access_connection_reference
  (-> | <-> ) destination_data_access_connection_reference
| refined to data access )
[ { { data_access_connection_property_association }+ } ]
[ in_modes_and_transitions ] ;

bus_access_connection ::=
bus_access_connection_identifier :
( bus access source_bus_access_connection_reference
  (-> | <-> ) destination_bus_access_connection_reference
| refined to bus access )
[ { { bus_access_connection_property_association }+ } ]
[ in_modes_and_transitions ] ;

subprogram_access_connection ::=
subprogram_access_connection_identifier :
( subprogram access
source_subprogram_access_connection_reference
(-> | <-> ) destination_subprogram_access_connection_reference
| refined to subprogram access )
[ { { subprogram_access_connection_property_association }+ } ]
[ in_modes_and_transitions ] ;

subprogram_group_access_connection ::=
subprogram_group_access_connection_identifier :
( subprogram group  access
source_subprogram_group_access_connection_reference
(-> | <-> )
destination_subprogram_group_access_connection_reference
| refined to subprogram group access )
[ { { data_access_connection_property_association }+ } ]
[ in_modes_and_transitions ] ;
```

continues

```
access_connection_reference ::=
   -- requires or provides access feature in the component type
   requires_access_identifier | provides_access_identifier |
   -- requires or provides access feature a feature group
   -- of the component type
   feature_group_identifier . requires_access_identifier |
   feature_group_identifier . provides_access_identifier |
   -- provides or requires access in a subcomponent
   subcomponent_identifier . provides_access_identifier |
   subcomponent_identifier . requires_access_identifier |
   -- provides or requires access in a subprogram call
   subprogram_call_identifier . provides_access_identifier |
   subprogram_call_identifier . requires_access_identifier |
   -- data, subprogram, subprogram group or
   --   bus subcomponent being accessed
   accessed_subcomponent_identifier |
   -- subprogram a processor being accessed
   processor . provides_subprogram_access_identifier

feature_group_connection ::=
feature_group_connection_identifier :
(  feature group source_feature_group_connection_reference   <->
destination_feature_group_connection_reference
| refined to feature group )
[ { { feature_group_connection_property_association }+ } ]
[ in_modes_and_transitions ] ;

feature_group_reference ::=
   -- feature group in the component type
   feature_group_identifier |
   -- feature group in a subcomponent
   subcomponent_identifier . feature_group_identifier |
   -- feature group element in a feature group
   -- of the component type
   feature_group_identifier . element_feature_group_identifier

-- call sequences within a thread or subprogram
subprogram_call_sequence ::=
call_sequence_identifier : { { subprogram_call }+ }
[ { { call_sequence_property_association }+ } ]
[ in_modes ] ;

subprogram_call ::=
subprogram_call_identifier : subprogram called_subprogram
[ { { subcomponent_call_property_association }+ } ] ;

called_subprogram ::=
     -- identification by classifier
     subprogram_component_classifier_reference |
     data_component_type_reference .
       data_provides_subprogram_access_identifier |
     subprogram_group_component_type_reference .
       provides_subprogram_access_identifier |
```

```
      feature_group_identifier .
         requires_subprogram_access_identifier |
      abstract_component_type_reference .
         requires_subprogram_access_identifier |
      -- identification by prototype
      subprogram_prototype_identifier |
      -- identification by processor subprogram access feature
      processor . provides_subprogram_access_identifier |
      -- identification by subprogram instance
      subprogram_subcomponent_identifier |
      subprogram_group_subcomponent_identifier .
         provides_subprogram_access_identifier |
      requires_subprogram_access_identifier |
      requires_subprogram_group_access_identifier .
         provides_subprogram_access_identifier

-- flow specifications from input to output
-- flow specification refinements must have
-- a property association or an in modes
flow_spec ::= flow_source_spec | flow_sink_spec | flow_path_spec

flow_source_spec ::=
flow_identifier :  ( flow source out_flow_feature_identifier
                   | refined to flow source )
[ { { property_association }+ } ]
[ in_modes_and_transitions ] ;

flow_sink_spec ::=
flow_identifier : ( flow sink in_flow_feature_identifier
                  | refined to flow sink )
[ { { property_association }+ } ]
[ in_modes_and_transitions ] ;

flow_path_spec ::=
flow_identifier :
(( flow path in_flow_feature_identifier ->
            out_flow_feature_identifier )
| refined to flow path )
[ { { property_association }+ } ]
[ in_modes_and_transitions ] ;

flow_feature_identifier ::=
feature_identifier |
 feature_group_identifier.feature_identifier |
 feature_group_identifier.feature_group_identifier

-- realization of a flow specification in the component implementation
-- flow implementations cannot be refined
flow_implementation ::=
flow_source_implementation | flow_sink_implementation |
    flow_path_implementation
[ { { property_association }+ } ]
[ in_modes_and_transitions ] ;
```

continues

```
flow_source_implementation ::=
flow_identifier : flow source
    { subcomponent_flow_reference -> connection_identifier -> }*
    out_flow_feature_identifier

flow_sink_implementation ::=
flow_identifier : flow sink flow_feature_identifier
    { -> connection_identifier -> subcomponent_flow_reference }*

flow_path_implementation ::=
flow_identifier : flow path in_flow_feature_identifier
    [ { -> connection_identifier -> subcomponent_flow_identifier }+
      -> connection_identifier ] -> out_flow_feature_identifier

-- flow through subcomponents and through data components
subcomponent_flow_reference::=
subcomponent_identifier [ . flow_spec_identifier ] |
 data_subcomponent_identifier |
 requires_data_access_identifier |
 provides_data_access_identifier

-- end-to-end flows (etef) can be composed of
-- other end-to-end flows and subcomponent flows
end_to_end_flow ::=
end_to_end_flow_identifier :
(( end to end flow start_subcomponent_flow_or_etef_identifier
 { -> connection_identifier ->
   flow_path_subcomponent_flow_or_etef_reference }*
  -> connection_identifier
  -> end_subcomponent_flow_or_etef_reference )
| refined to end to end flow )
[ { ( property_association }+ } ]
[ in_modes_and_transitions ] ;

subcomponent_flow_or_etef_reference ::=
subcomponent_flow_identifier | end_to_end_flow_identifier

-- modes and mode transitions
mode ::=
mode_identifier :  [ initial ] mode
[ { { mode_property_assocation }+ } ] ;

mode_transition ::=
[ mode_identifier : ] source_mode_identifier
    -[ mode_transition_trigger { , mode_transition_trigger }* ]->
    destination_mode_identifier
    [ { { mode_transition_property_assocation }+ } ] ;

mode_transition_trigger ::=
    port_identifier | subcomponent_identifier . port_identifier |
    self . event_source_identifier | processor . port_identifier
```

```
-- mode-specific subcomponents, connections, flows, annexes
in_modes ::=
in modes ( ( mode_identifier { , mode_identifier }*  ) )

component_in_modes ::=
in modes ( ( mapped_mode_name { , mapped_mode_name }*  ) )

mapped_mode_name ::=
    mode_identifier [ => subcomponent_mode_identifier ]

in_modes_and_transitions ::=
in modes ( ( mode_or_transition { , mode_or_transition }* ) )

mode_or_transition ::= mode_identifier | mode_transition_identifier

-- name-binding of inheritable modes
component_in_modes ::=
    in modes ( ( mode_name { , mode_name }* ) | all )

-- optionally map mode identifier to a subcomponent mode identifier
mode_name ::=
local_mode_identifier [ => subcomponent_mode_identifier ]

-- sublanguage annotation
annex_subclause ::=
annex annex_identifier
 ( {** annex_specific_language_constructs **} | none )
 [ in_modes ] ;

annex_library ::=
annex annex_identifier
 ( {** annex_specific_reusable_constructs **} | none )
 [ in_modes ];

-- parameterization of component types and implementations
prototype ::=
    prototype_identifier : [ refined to ]
        ( component_prototype | feature_group_type_prototype
        | feature_prototype )
        [ { { prototype_property_association }+ } ] ;

component_prototype ::=
component_category [ component_classifier_reference ] [ [] ]

feature_group_type_prototype ::=
    feature group [ feature_group_type_reference ]

feature_prototype ::=
    [ in | out ] feature [ component_classifier_reference ]

prototype_bindings ::=
    ( prototype_binding { , prototype_ binding }* )
```

continues

```
prototype_binding ::=
    prototype_identifier =>
    ( component_prototype_actual
    | component_prototype_actual_list
    | feature_group_type_prototype_actual
    | feature_prototype_actual )

component_prototype_actual ::=
    component_category
    ( component_classifier_reference [ prototype_bindings ]
    | prototype_identifier )

component_prototype_actual_list ::=
    ( component_prototype_actual
      { , component_prototype_actual }* )

feature_group_type_prototype_actual ::=
        ( feature group feature_group_type_reference
          [ prototype_binding ] )
        | (feature group feature_group_type_prototype_identifier )

feature_prototype_actual ::=
    ( ( in | out | in out) ( event | data | event data ) port )
    [ component_classifier_reference ] )
    | ( ( requires | provides )
    ( bus | data | subprogram group | subprogram ) access
    [ component_classifier_reference ] )
    | ( [ in | out ] feature feature_prototype_identifier )

-- Assigning property values
property_association ::=
[ property_set_identifier :: ] property_name_identifier
 ( => | +=> )    [ constant ] assignment [ in_binding ] ;

contained_property_association ::=
[ property_set_identifier :: ] property_name_identifier
 ( => | +=> )  [ constant ] assignment
   applies to contained_model_path_element_identifier
     { . contained_ model_path_element _identifier }*
[ in_binding ] ;

-- path to model element to which property association applies.
-- Any named model element can be part of path
contained_model_element_path ::=
    ( contained_model_element { . contained_model_element }*
      [ annex_path ] )
    | annex_path

contained_model_element ::=
    named_element_identifier
    | named_element_array_selection_identifier
```

```
-- path leading to model elements in an annex subclause
annex_path ::=
    annex annex_identifier {** <annex specific path> **}

-- deployment binding specific property values
-- (e.g., processor type specific execution time).
-- Platform components are processor, bus, memory.
in_binding ::=
in binding ( platform_component_classifier_reference
    { , platform_component_classifier_reference }* )

-- values(s) to be assigned
assignment ::= property_value | modal_property_value

modal_property_value ::=
 property_value in_modes , }* property_value [ in_modes ]

property_value ::= property_expression | property_value_list

-- If the property expects a list then a single value
-- must be enclosed in "()"
property_value_list ::=
 ( [ ( property_expression | property_value_list )
    { , ( property_expression | property_value_list ) }* ] )

-- single property value
property_expression ::=
 boolean_term | real_term | integer_term | string_term |
 enumeration_term | real_range_term | integer_range_term |
 property_term | component_classifier_term | reference_term |
 record_term | computed_term

boolean_term ::= true | false

real_term ::= signed_aadlreal_or_constant

integer_term ::= signed_aadlinteger_or_constant

string_term ::= string_literal | string_property_constant_term

enumeration_term ::=
 enumeration_identifier | enumeration_property_constant_term

integer_range_term ::=
 integer_term .. integer_term [ delta integer_term ] |
    integer_range_property_constant_term

real_range_term ::=
 real_term .. real_term [ delta real_term ] |
    real_range_property_constant_term
```

continues

```
property_term ::=
 [ property_set_identifier :: ] property_identifier

property_constant_term ::=
 [ property_set_identifier :: ] property_constant_identifier

component_classifier_term ::=
 classifier ( component_classifier_reference )

reference_term ::=
 reference ( contained_model_element_path )

record_term ::=
 [ record_field_identifier => property_value ;
     { record_field_identifier => property_value ; }*  ]

computed_term ::= compute ( function_identifier )

-- Defining new properties
property_set ::=
    property set property_set_identifier is
        { import_declaration }*
        { property_type_definition | property_definition
        | property_constant_definition }*
    end property_set_identifier ;

--   property type is defined in terms of a base type,
-- inline type constructor , or type reference
property_type_definition ::=
    property_type_identifier : type basic_property_type
      | property_type_constructor ;

-- built-in basic types and type constructors
basic_property_type ::=
 aadlboolean | aadlstring | aadlreal | aadlinteger

property_type_constructor ::=
 enumeration_type | units_type | number_type | range_type
 | classifier_type | reference_type

enumeration_type ::=
 enumeration ( enumeration_literal_identifier
     { , enumeration_literal_identifier }* )

units_type ::=
 units ( defining_unit_identifier
     { , defining_unit_identifier =>
         unit_identifier * numeric_literal }* )

number_type ::=
 aadlreal units units_designator  | aadlreal real_range |
  aadlreal real_range units units_designator |
  aadlinteger units units_designator |
```

```
    aadlinteger integer_range |
    aadlinteger integer_range units units_designator|

units_designator ::= units_property_type_reference | units_list

-- integer and real ranges
real_range ::=
 signed_aadlreal_or_constant .. signed_aadlreal_or_constant

integer_range ::=
 signed_aadlreal_or_constant .. signed_aadlreal_or_constant

signed_aadlreal_or_constant ::=
[ + | - ] integer_literal [ unit_identifier ] |
 [ + | - ] real_property_constant_term

signed_aadlinteger_or_constant ::=
([ + | - ] real_literal [ unit_identifier ] |
[ + | - ] integer_property_constant_term )

range_type ::=
 range of aadlreal | range of aadlinteger |
 range of number_property_type_reference

classifier_type ::=
 classifier [ ( classifier_category_reference
     { , classifier_category_reference }* ) ]

-- any Meta model element that is a subclass of classifier
classifier_category_reference ::=
 [ { annex_identifier }** ] meta_model_class_identifier

reference_type ::=
 reference [ ( named_element_meta_model_identifier
        { , named_element_meta_model_identifier }* ) ]

record_type ::=
  record ( { field_identifier :
                { list of }* property_type_reference ; }+ )

-- all properties other than predeclared properties
-- must be qualified by their property set
property_type_reference ::=
 [ property_set_identifier :: ] property_type_identifier

property_type_designator ::=
 basic_property_type | property_type_constructor
 | property_type_reference

-- new user defined property
property_definition ::= property_identifier : [ inherit ]
    ( single_valued_property | multi_valued_property )
```

continues

```
applies to ( ( property_owner { , property_owner }*
                  | all ) ) ;

single_valued_property ::= property_type_designator
    [ => default_property_expression ]

multi_valued_property ::= { list of }+ property_type_designator
    [ => ( [ default_property_expression
            { , default_property_expression }* ] ) ]

property_owner ::=
    named_element_meta_model_identifier |
    component_classifier_reference |
    feature_group_type_reference
-- property constant with optional default value.
property_constant ::=
    property_constant_identifier : constant
    property_type_designator => property_value ;
```

A.2 Component Type and Implementation Elements

Table A-1 contains a summary of the elements allowed in component types and implementations in AADL.

Table A-1: *Constraints/Restrictions for Application Software Components*

Component Category	Type	Implementation
Data	Features: • Provides subprogram access • Requires subprogram access • Requires subprogram group access • Feature group • Feature Flow specifications: No Modes: No Properties: Yes	Subcomponents: • Data • Subprogram • Abstract Subprogram calls: No Connections: Yes Flows: No Modes: No Properties: Yes

Table A-1: *Constraints/Restrictions for Application Software Components (continued)*

Component Category	Type	Implementation
Subprogram	Features: • Out event port • Out event data port • Feature group • Requires data access • Requires subprogram access • Requires subprogram group access • Parameter • Feature Flow specifications: Yes Modes: Yes Properties: Yes	Subcomponents: • Data • Subprogram • Abstract Subprogram calls: Yes Connections: Yes Flows: Yes Modes: Yes Properties: Yes
Subprogram Group	Features: • Feature group • Provides subprogram access • Requires subprogram access • Requires subprogram group access • Provides subprogram group access • Feature Flow specification: No Modes: No Properties: Yes	Subcomponents: • Data • Subprogram • Subprogram group • Abstract Subprogram calls: No Connections: Yes Flows: No Modes: No Properties: Yes
Thread	Features: • Provides subprogram access • Requires subprogram access • Provides subprogram group access • Requires subprogram group access • Port • Feature group • Requires data access • Feature Flow specifications: Yes Modes: Yes Properties: Yes	Subcomponents: • Data • Subprogram • Subprogram group • Abstract Subprogram calls: Yes Connections: Yes Flows: Yes Modes: Yes Properties: Yes

continues

Table A-1: *Constraints/Restrictions for Application Software Components (continued)*

Component Category	Type	Implementation
Thread Group	Features: • Provides subprogram access • Requires subprogram access • Provides subprogram group access • Requires subprogram group access • Port • Feature group • Provides data access • Requires data access • Feature Flow specifications: Yes Modes: Yes Properties: Yes	Subcomponents: • Data • Subprogram • Subprogram group • Thread • Thread group • Abstract Subprogram calls: No Connections: Yes Flows: Yes Modes: Yes Properties: Yes
Process	Features: • Provides subprogram access • Requires subprogram access • Provides subprogram group access • Requires subprogram group access • Port • Feature group • Provides data access • Requires data access • Feature Flow specifications: Yes Modes: Yes Properties: Yes	Subcomponents: • Data • Subprogram • Subprogram group • Thread • Thread group • Abstract Subprogram calls: No Connections: Yes Flows: Yes Modes: Yes Properties: Yes
Processor	Features: • Provides subprogram access • Provides subprogram group access • Port • Feature group • Requires bus access • Provides bus access • Feature Flow specifications: Yes Modes: Yes Properties: Yes	Subcomponents: • Memory • Bus • Virtual processor • Virtual bus • Abstract Subprogram calls: No Connections: Yes Flows: Yes Modes: Yes Properties: Yes

Table A-1: *Constraints/Restrictions for Application Software Components (continued)*

Component Category	Type	Implementation
Virtual Processor	Features: • Provides subprogram access • Provides subprogram group access • Port • Feature group • Feature Flow specification: Yes Modes: Yes Properties: Yes	Subcomponents: • Virtual processor • Virtual bus • Abstract Subprogram calls: No Connections: Yes Flows: Yes Modes: Yes Properties: Yes
Memory	Features: • Requires bus access • Provides bus access • Feature group • Feature Flow specifications: No Modes: Yes Properties: Yes	Subcomponents: • Memory • Bus • Abstract Subprogram calls: No Connections: Yes Flows: No Modes: Yes Properties: Yes
Bus	Features: • Requires bus access • Feature group • Feature Flow specifications: No Modes: Yes Properties: Yes	Subcomponents: • Virtual bus • Abstract Subprogram calls: No Connections: No Flows: No Modes: Yes Properties: Yes
Virtual Bus	Features: • None Flow specifications: No Modes: Yes Properties: Yes	Subcomponents: • Virtual bus • Abstract Subprogram calls: No Connections: No Flows: No Modes: Yes Properties: Yes

continues

Table A-1: *Constraints/Restrictions for Application Software Components (continued)*

Component Category	Type	Implementation
Device	Features: • Port • Feature group • Provides subprogram access • Provides subprogram group access • Requires bus access • Provides bus access • Feature Flow specifications: Yes Modes: Yes Properties: Yes	Subcomponents: • Bus • Virtual bus • Data • Abstract Subprogram calls: No Connections: Yes Flows: Yes Modes: Yes Properties: Yes
System	Features: • Provides subprogram access • Requires subprogram access • Provides subprogram group access • Requires subprogram group access • Port • Feature group • Provides data access • Provides bus access • Requires data access • Requires bus access • Feature Flow specifications: Yes Modes: Yes Properties: Yes	Subcomponents: • Data • Subprogram • Subprogram group • Process • Processor • Virtual processor • Memory • Bus • Virtual bus • Device • System • Abstract Subprogram calls: No Connections: Yes Flows: Yes Modes: Yes Properties: Yes

Table A-1: *Constraints/Restrictions for Application Software Components (continued)*

Component Category	Type	Implementation
Abstract	Features: • Port • Feature group • Provides data access • Requires data access • Provides subprogram access • Requires subprogram access • Provides subprogram group access • Requires subprogram group access • Provides bus access • Requires bus access • Feature Flow specification: Yes Modes: Yes Properties: Yes	Subcomponents: • Data • Subprogram • Subprogram group • Thread • Thread group • Process • Processor • Virtual processor • Memory • Bus • Virtual bus • Device • System • Abstract Subprogram calls: Yes Connections: Yes Flows: Yes Modes: Yes Properties: Yes

A.3 Basic Property Types and Type Constructors

Table A-2 summarizes the basic property types and property type constructors provided by the AADL standard [AS-5506A].

Table A-2: *AADL Property Types*

Property Type	Definition
aadlboolean	Boolean values: true or false
aadlstring	String value: ASCII characters between two quotation marks
enumeration	Literal value: literal from an ordered list of enumeration literals

continues

Table A-2: *AADL Property Types (continued)*

Property Type	Definition
units	Units value: unit literal from a list of measurement unit identifiers with conversion factors
aadlreal	Real value or a real value and its measurement unit
aadlinteger	Integer value or an integer value and its measurement unit
range of	Real or integer range value: represents a closed interval of numbers by specifying its lower and upper bound
classifier	Classifier reference value: any classifier whose Meta model class matches that of the property type definition
reference	Model element reference value: named path to any model element whose Meta model class matches that of the property type definition
record	A collection of typed property values, each identified by name (record field)

A.4 AADL Reserved Words

Table A-3 lists the AADL reserved words. Reserved words are case insensitive.

Table A-3: *AADL Reserved Words*

aadlboolean	applies	delta	flow
aadlinteger	binding	device	flows
aadlreal	bus	end	group
aadlstring	calls	enumeration	implementation
abstract	classifier	event	in
access	compute	extends	inherit
all	connections	false	initial
and	constant	feature	inverse
annex	data	features	is

Table A-3: *AADL Reserved Words (continued)*

list	path	range	subcomponents
memory	port	record	subprogram
mode	private	reference	system
modes	process	refined	thread
none	processor	renames	to
not	properties	requires	true
of	property	self	type
or	prototypes	set	units
out	provides	sink	virtual
package	public	source	with
parameter			

A.5 AADL Properties

The next set of tables provides a summary of AADL properties and property types that have been predeclared in the SAE AADL Standard document [AS5506A]. These properties have been organized into several property sets. However, you do not have to qualify the property name with the property set name when using it.

The set of predeclared property sets *Deployment_Properties, Thread_ Properties, Timing_Properties, Communication_Properties, Memory_Properties, Programming_Properties*, and *Modeling_Properties* is part of every AADL specification. This set defines properties for AADL model elements that are defined in the core of the AADL. A modeler may not modify these property sets. *Deployment_Properties* contains properties related to the deployment of the embedded application on the execution platform. *Thread_Properties* contains properties that characterize threads and their features. *Timing_Properties* contains properties related to execution timing. *Communication_Properties* contains properties that describe the nature of communication such as topology, protocol, and queuing characteristics. *Memory_Properties* contains properties related to memory as storage, data access, and device access. *Programming_ Properties* contains properties for relating AADL models to application programs. *Modeling_Properties* contains properties related to the AADL model itself.

The property set *AADL_Project* defines property enumeration types and property constants that can be tailored for different AADL projects and site installations.

A.5.1 Deployment Properties

Table A-4 summarizes properties associated with binding software, data, and connections to execution platform components and related deployment properties. The table includes property declarations taken from the standard and examples of property associations. Since all standard and user-defined properties that can be specified for at least one component of any category can be specified for an abstract component, the abstract component category is not included in the table.

Table A-4: *Deployment Properties*

Property	Description
Actual_Connection_ Binding	An inherited property that binds connections and virtual buses to a list of one or more processor, virtual processor, bus, virtual bus, device, or memory components. It can be specified for features, connections, threads, thread groups, processes, systems, and virtual buses. The entries in the list represent the flow sequence of the connection through the execution platform. It is a property of type **list of reference**. Declaration: Actual_Connection_Binding: **inherit list of reference** (processor, virtual processor, bus, virtual bus, device, memory) **applies to** (feature, connection, thread, thread group, process, system, virtual bus); Property Association Examples: DC: **data port** *speed_control.command_data -> throttle.cmd {Actual_Con- nection_Binding =>* (**reference**(*Hi_speed_bus*));}; *Actual_Connection_Binding =>* (**reference**(*Hi_speed_bus*), **reference**(*global_proc*));

Table A-4: *Deployment Properties (continued)*

Property	Description
Allowed_Connection_Binding	An inherited property that constrains the binding of a connection or virtual bus to a list of one or more processor, virtual processor, bus, virtual bus, device, or memory components. It can be specified for features, connections, threads, thread groups, processes, systems, and virtual buses. When specified for a feature such as a port it indicates a binding constraint for all connections through that feature. It is a property of type **list of reference**. Declaration: Allowed_Connection_Binding: **inherit list of reference** (processor, virtual processor, bus, virtual bus, device, memory) **applies to** (feature, connection, thread, thread group, process, system, virtual bus); Property Association Examples: *Allowed_Connection_Binding => (**reference**(Hi_speed_VBbus)) ;* *EC: **event port** interface_unit.disengage -> speed_control.disengage {Allowed_Connection_Binding => (**reference**(Hi_speed_bus), **reference**(Lo_speed_bus)) ;}*
Allowed_Connection_Binding_Class	An inherited property that constrains the binding of a connection or virtual bus to a list of one or more processor, virtual processor, bus, virtual bus, device, or memory classifiers. It can be specified for features, connections, threads, thread groups, processes, systems, and virtual buses. This property constrains binding to components that are consistent with the set of classifiers listed. It is a property of type **list of classifiers**. Declaration: Allowed_Connection_Binding_Class: **inherit list of classifier** (processor, virtual processor, bus, virtual bus, device, memory) **applies to** (feature, connection, thread, thread group, process, system, virtual bus); Property Association Examples: *Allowed_Connection_Binding_Class => (**classifier**(basic), **classifier**(basic2), **classifier**(interface, **classifier**(v_bus));* *Allowed_Connection_Binding_Class => (**classifier**(basic));*

continues

Table A-4: *Deployment Properties (continued)*

Property	Description
Actual_Memory_ Binding	An inherited property that binds code and data produced from source text to a list of one or more memory components. It can be specified for threads, thread groups, processes, systems, processors, devices, data, data ports, event data ports, subprograms, and subprogram groups, and is a property of type **list of reference**. Declaration: Actual_Memory_Binding: **inherit list of reference** (memory) **applies to** (thread, thread group, process, system, processor, device, data, data port, event data port, subprogram, subprogram group); Property Association Examples: *Actual_Memory_Binding => (**reference**(Standard_RAM));* *Actual_Memory_Binding =>(**reference** (Hi_speed_RAM), **reference**(RT_Cache));*
Allowed_Memory_ Binding	An inherited property that binds code and data produced from source text to a list of one or more memory, system, or processor components. It can be specified for threads, thread groups, processes, systems, processors, devices, data, data ports, event data ports, subprograms, and subprogram groups. A single name may be used to specify the binding to the named memory, thereby defining the actual binding. When specified for a processor or system component, it constrains the binding to memory contained in that component. It is a property of type **list of reference**. Declaration: Allowed_Memory_Binding: **inherit list of reference** (memory, system, processor) **applies to** (thread, thread group, process, system, device, data, data port, event data port, subprogram, subprogram groups, processor); Property Association Examples: *Allowed_Memory_Binding => (**reference**(RT_RAM));* *Allowed_Memory_Binding => (**reference**(Standard_RAM), **reference**(Hi_Speed_Ram));*

Table A-4: *Deployment Properties (continued)*

Property	Description
Allowed_Memory_Binding_Class	An inherited property that constrains the *Allowed_Memory_Binding* property to components that are consistent with a list of one or more memory, processor, and system classifiers. The constraint is such that memory components that can be used for the property *Allowed_Memory_Binding* must be instances of one of the memory classifiers listed or be a memory component contained in an instance of one of the processor or system classifiers listed. It can be specified for threads, thread groups, processes, systems, devices, data, data ports, event data ports, subprograms, subprogram groups, and processors. If no value is specified for *Allowed_Memory_Binding_Class* property, no constraints are imposed on the *Allowed_Memory_Binding* property. Declaration: Allowed_Memory_Binding_Class: **inherit list of classifier** (memory, system, processor) **applies to** (thread, thread group, process, system, device, data, data port, event data port, subprogram, subprogram group, processor); Property Association Examples: *Allowed_Memory_Binding_Class => (classifier(RT_RAM), classifier(RT.basic_3GHz)) ;* *Allowed_Memory_Binding => (classifier(core_sys));*
Actual_Processor_Binding	An inherited property that binds threads, device drivers, or virtual processors to a list of one or more processor or virtual processor components. If multiple processors are listed (e.g., multicore), a scheduler can be used to dynamically allocate the threads or drivers. It can be specified for threads, thread groups, processes, systems, virtual processors, and devices. Binding a device to a processor is the binding of the device's driver software to the processor. In binding a virtual processor to a processor, the threads, device drivers, and other virtual processors bound to that virtual processor are then bound to the processor component. Declaration: Actual_Processor_Binding: **inherit list of reference** (processor, virtual processor) **applies to** (thread, thread group, process, system, virtual processor, device); Property Association Examples: *Actual_Processor_Binding => (reference(Standard_Processor));* *Actual_Processor_Binding => (reference(standard_processor), reference(basic_proc));*

continues

Table A-4: *Deployment Properties (continued)*

Property	Description
Allowed_Processor_ Binding	An inherited property that constrains the binding of threads, device drivers, or virtual processors to a list of one or more processors, virtual processor or system components. Systems included in the list represent only the virtual processors and processors contained within those systems. It can be specified for threads, thread groups, processes, systems, virtual processors, or devices. When specified for a thread or virtual processor, the thread or virtual processor can be bound to any of the listed processors or virtual processors. When specified for a thread group, process, or system, it applies to all contained threads unless explicitly overridden. When specified for a device, it constrains the binding of the device's driver software. Declaration: Allowed_Processor_Binding: **inherit list of reference** (processor, virtual processor, system) **applies to** (thread, thread group, process, system, virtual processor, device); Property Association Examples: *Allowed_Processor_Binding => (**reference**(RT_proc)) ;* *Allowed_Processor_Binding => (**reference**(Basic_proc), **reference**(RT_3GHz)) ;*
Allowed_Processor_ Binding_Class	An inherited property that constrains the binding of threads, device drivers, or virtual processors to a list of one or more processors, virtual processor, or systems classifiers. This property constrains the binding to processor or virtual processor components that are instances of the processor or virtual processor classifiers listed in its property association. If a system classifier is listed, the binding is constrained to processor and virtual processor components contained within instances of the system. It can be specified for threads, thread groups, processes, systems, virtual processors, and devices. When specified for a thread or virtual processor, the thread or virtual processor can be bound to instances of any of the listed processors or virtual processors. When specified for a thread group, process, or system, it applies to all contained threads unless explicitly overridden. When specified for a device, it constrains the binding of the device's driver software. Declaration: Allowed_Processor_Binding_Class: **inherit list of classifier** (processor, virtual processor, system) **applies to** (thread, thread group, process, system, virtual processor, device);

Table A-4: *Deployment Properties (continued)*

Property	Description
Allowed_Processor_ Binding_Class (continued)	Property Association Examples: *Allowed_Processor_Binding_Class =>(**classifier**(Proc_sys.impl), **classifier**(VP_basic)) ;*
Actual_Subprogram_ Call	A property that specifies the subprogram instance whose code is servicing a subprogram call. It can be specified only for subprogram accesses. The subprogram instance may be in other threads (remote subprograms) or local. For local subprograms, the property identifies a specific local code instance. If no value is specified, the subprogram call is a local call. This property can be used to model sharing of subprogram. Declaration: Actual_Subprogram_Call: **reference** (subprogram) **applies to** (subprogram access); Property Association Example: *Actual_Subprogram_Call => **reference**(get_value);*
Allowed_Subprogram_ Call	A property that constrains the binding of a subprogram call to a list of one or more subprogram instances. It can be specified only for subprogram accesses. The subprogram instances may be in other threads (remote subprograms) or local. For local subprograms, the property identifies a set of local code instances. If no value is specified, the subprogram calls are local. Declaration: Allowed_Subprogram_Call: **list of reference** (subprogram) **applies to** (subprogram access); Property Association Examples: *Allowed_Subprogram_Call => (**reference**(calc_value));* *Allowed_Subprogram_Call => (**reference**(get_temp), **reference**(get_press));* *Allowed_Subprogram_Call => (**reference**(total_all), **reference**(calc_press)) **applies to** calculations.service_access;*

continues

Table A-4: *Deployment Properties (continued)*

Property	Description
Actual_Subprogram_Call_Binding	An inherited property that binds a remote subprogram call to a list of one or more bus, processor, memory, or device components. It can be specified only for a subprogram. If a value is not specified, the subprogram call is a local call or the binding is inferred from the bindings of the caller and callee. Declaration: Actual_Subprogram_Call_Binding: **inherit list of reference** (bus, processor, memory, device) **applies to** (subprogram); Property Association Examples: *Actual_Subprogram_Call_Binding: => (**reference**(my_mem), **reference**(my_proc), **reference**(my_bus)) ;*
Allowed_Subprogram_Call_Binding	An inherited property that constrains the binding of a remote subprogram call to a list of one or more bus, processor, or device components. It can be specified for subprograms, threads, thread groups, processes, or systems. If a value is not specified, the subprogram call may be a local call or the binding is inferred from the bindings of the caller and callee. Declaration: Allowed_Subprogram_Call_Binding: **inherit list of reference** (bus, processor, device) **applies to** (subprogram, thread, thread group, process, system); Property Association Example: *Allowed_Subprogram_Call_Binding => (**reference**(basic_proc));*
Not_Collocated	A property that constrains the use of hardware resources such that hardware resources used by several software components must be distinct. It can be specified for processes and systems. This is a **record** property type consisting of a *Targets* field (list of one or more components) and a *Location* field (a single classifier). The Targets are a list of one or more data, thread, process, system components, or connections that are not to be co-located on the same instance of a specific processor, memory, bus, or system classifier (the *Location* field). In summary, the components referenced by Targets must not be bound to the same hardware of the type specified in the Location field. If the Location is a system component, they may not be collocated on any component contained in the system component.

Table A-4: *Deployment Properties (continued)*

Property	Description
Not_Collocated (continued)	Declaration: Not_Collocated: **record** (Targets: **list of reference** (data, thread, process, system, connection); Location: **classifier** (processor, memory, bus, system);) **applies to** (process, system); Property Association Example: *Not_Collocated => [Targets => (**reference**(control_1), **reference**(control_2)) ; Location => **classifier**(Basic);];*
Collocated	A property that constrains the use of hardware resources such that several software components must be bound to the same hardware component. It can be specified for processes and systems. This is a **record** property type consisting of a *Targets* field (list of components) and a *Location* field (a classifier). The Targets are a list of one or more data, thread, process, system components, or connections that are to be co-located on the same instance of a specific processor, memory, bus, or system classifier (the *Location* field). In summary, the components referenced by *Targets* must be bound to the same hardware component that is an instance of that specified in the *Location* field. If the *Location* is a system classifier, the *Target* components are to be collocated on a (any) single component contained in an instance of the system classifier. Declaration: Collocated: **record** (Targets: **list of reference** (data, thread, process, system, connection); Location: **classifier** (processor, memory, bus, system);) **applies to** (process, system); Property Association Example: *Collocated => [Targets => (**reference**(control_1), **reference**(control_2)) ; Location => **classifier**(Basic.M3);];*

continues

Table A-4: *Deployment Properties (continued)*

Property	Description
Provided_Virtual_Bus_ Class	An inherited property that specifies a list of one or more virtual bus classifiers (protocols) supported by a bus, virtual bus, virtual processor, device, or processor. It can be specified for buses, virtual buses, processors, virtual processors, devices, memory, and systems. The property indicates that a component, with a binding requirement for a virtual bus classifier, can be bound to a component whose *Provided_Virtual_Bus_Class* value includes the desired virtual bus classifier. It is not required that a component with this property have a virtual bus subcomponent. Declaration: Provided_Virtual_Bus_Class : **inherit list of classifier** (virtual bus) **applies to** (bus, virtual bus, processor, virtual processor, device, memory, system); Property Association Example: *Provided_Virtual_Bus_Class* => (**classifier**(*virtual_hispd.impl*));
Required_Virtual_Bus_ Class	An inherited property that specifies a list of one or more virtual bus classifiers (protocols) to which a connection or virtual bus must be bound. This is such that the binding must be to one instance of each of the specified classifiers. This property can be specified for virtual buses, connections, ports, threads, thread groups, processes, systems, and devices. Declaration: Required_Virtual_Bus_Class : **inherit list of classifier** (virtual bus) **applies to** (virtual bus, connection, port, thread, thread group, process, system, device); Property Association Example: *Required_Virtual_Bus_Class* => (**classifier**(*virtual_hispd.impl*));
Provided_Connection_ Quality_Of_Service	An inherited property that specifies a list of one or more quality of service values provided by a protocol supported by a virtual bus, bus, virtual processor, or processor. This property can be specified for virtual buses, buses, virtual processors, processors, systems, devices, or memory. It is a property of type **list of** *Supported_Connection_QoS*. The type *Supported_Connection_QoS* is declared in the *AADL_project* property set with the values *GuaranteedDelivery*, *OrderedDelivery*, and *SecureDelivery*. Other values are *project-specified* that can be specified by a user.

Table A-4: *Deployment Properties (continued)*

Property	Description
Provided_Connection_ Quality_Of_Service (*continued*)	Declaration: Provided_Connection_Quality_Of_Service : **inherit list of** Supported_Connection_QoS **applies to** (bus, virtual bus, processor, virtual processor, system, device, memory); Property Association Examples: *Provided_Connection_Quality_Of_Service => (GuaranteedDelivery, SecureDelivery);* *Provided_Connection_Quality_Of_Service => (OrderedDelivery);*
Required_Connection_ Quality_Of_Service	An inherited property that specifies a list of one or more quality of service values that a connection or virtual bus expects from a transmission protocol. This property can be specified for ports, connections, virtual buses, threads, thread groups, processes, systems, and devices, and is a property of type **list of** *Supported_Connection_QoS*. The type *Supported_Connection_QoS* is declared in the *AADL_project* property set with the values *GuaranteedDelivery*, *OrderedDelivery*, and *SecureDelivery*. Other values are *project-specified* that can be specified by a user. Declaration: Required_Connection_Quality_Of_Service : **inherit list of** Supported_Connection_QoS **applies to** (port, connection, virtual bus, thread, thread group, process, system, device); Property Association Examples: *Required_Connection_Quality_Of_Service => (GuaranteedDelivery, SecureDelivery);* *Required_Connection_Quality_Of_Service => (OrderedDelivery) ;*
Allowed_Connection_ Type	A property that specifies a list of one or more connections supported by a bus, such that a connection may only be legally bound to a bus if the bus supports that category of connection. It can be specified for buses and devices. The connections are an enumerated list that includes *Sampled_Data_Connection, Immediate_Data_Connection, Delayed_Data_Connection, Port_Connection, Data_Access_Connection, or Subprogram_Access_Connection*. If a list of allowed connection protocols is not specified for a bus, any category of connection can be bound to the bus. If specified for a device, the property applies to all buses contained within that device and made accessible externally.

continues

Table A-4: *Deployment Properties (continued)*

Property	Description
Allowed_Connection_Type *(continued)*	Declaration: Allowed_Connection_Type: **list of enumeration** (Sampled_Data_ Connection, Immediate_Data_Connection, Delayed_Data_Connection, Port_Connection, Data_Access_Connection, Subprogram_Access_Connection) **applies to** (bus, device); Property Association Examples: *Allowed_Connection_Type = (Sampled_Data_Connection, Immediate_Data_Connection);* *Allowed_Connection_Type =>(Sampled_Data_Connection);*
Allowed_Dispatch_ Protocol	A property that specifies a list of one or more thread dispatch protocols supported by a processor. This is a property of type **list of** *Supported_Dispatch_Protocols*. *Supported_Dispatch_Protocols* is defined in the *AADL_project* property set. The property *Allowed_Dispatch_Protocol* can be specified only for processors and virtual processors. The standard defines the value *Periodic* for this property type. The other values are *project-specified* that can be specified by a user. A thread can only be legally bound to a processor if the thread dispatch protocol of the processor corresponds to the dispatch protocol required by the thread. If a list of allowed scheduling protocols is not specified for a processor, a thread with any dispatch protocol can be bound to and executed by the processor. Declaration: Allowed_Dispatch_Protocol: **list of** Supported_Dispatch_Protocols **applies to** (processor, virtual processor); Property Association Examples (assumes *Sporadic* and *Aperiodic* have been defined in *AADL_project*): *Allowed_Dispatch_Protocol = (Periodic);* *Allowed_Dispatch_Protocol =>(Periodic, Sporadic, Aperiodic);*
Allowed_Period	The property specifies a list of one or more allowed periods for periodic tasks bound to a processor. It can be specified for processors, virtual processors, and systems and is of type **list of** *Time_Range*. Each entry in the list specifies a range of *Time* values that are expressed using the standard units of ps (picoseconds), ns (nanoseconds), us (microseconds), ms (milliseconds), sec (seconds), min (minutes), and hr (hours). The period of every thread bound to the processor must fall within one of the specified ranges. If an allowed period is not specified for a processor, there are no restrictions on the periods of threads bound to that processor.

Table A-4: *Deployment Properties (continued)*

Property	Description
Allowed_Period *(continued)*	Declaration: Allowed_Period: **list of** Time_Range **applies to** (processor, system, virtual processor); Property Association Examples: Allowed_Period = *(20 ms … 40 ms, 100 ms.. 150 ms);* Allowed_Period =>*(20 ms … 100 ms);*
Allowed_Physical_ Access_Class	A property that specifies a list of one or more processors, devices, memory, and buses classifiers, whose instances are allowed to be connected to a bus. The property can be specified only for buses. If this property is not specified for a bus, the bus may be used to connect devices and memory to a processor. Declaration: Allowed_Physical_Access_Class: **list of classifier** (device, processor, memory, bus) **applies to** (bus); Property Association Example: *Allowed_Physical_Access_Class => (**classifier**(CRAM.Hi_Speed)) ;*
Allowed_Physical_ Access	A property that specifies a list of one or more processors, devices, memory, and bus components that are allowed to be connected to a bus. The property can be specified only for buses. If the property is not specified for a bus, the bus may be connected to any device, memory, processor, or bus component. Declaration: Allowed_Physical_Access: **list of reference** (device, processor, memory, bus) **applies to** (bus); Property Association Example: *Allowed_Physical_Access => (**reference**(cache_memory)) ;*

continues

Table A-4: *Deployment Properties (continued)*

Property	Description
Memory_Protocol	A property that specifies memory access and storage behaviors and restrictions. It can be specified only for memory and is of type **enumeration** with values *read_only, write_only, read_write,* and *execute_only*. The default value is *read_write*. Writeable data produced from software source text may only be bound to memory components that have the *write_only* or *read_write* value for this property. Declaration: Memory_Protocol: **enumeration** (read_only, write_only, read_write, execute_only) => read_write **applies to** (memory); Property Association Example: *Memory_Protocol => read_write ;*
Scheduling_Protocol	An inherited property that specifies a list of one or more scheduling protocols supported by a processor's or a virtual processor's thread scheduler. This property can be specified only for virtual processors and processors. It is of type **list of** *Supported_Scheduling_Protocols*. *Supported_Scheduling_Protocols* is defined in the *AADL_project* property set. The standard does not prescribe a particular scheduling protocol and the values for a property of the type *Supported_Scheduling_Protocols* are *project specified* (i.e., can be declared by a user). Declaration: Scheduling_Protocol: **inherit list of** Supported_Scheduling_Protocols **applies to** (virtual processor, processor); Property Association Examples (assumes *RMS, EDF, and FixedTimeline* have been defined in *AADL_project*): *Scheduling_Protocol = (RMS, EDF, FixedTimeline);* *Scheduling_Protocol = (RMS);*
Preemptive_Scheduler	A property that specifies whether a processor can preempt a thread during its execution. Its possible values are **true** or **false**. If this property is not specified for a processor, its value is **true** (i.e., the processor has a preemptive scheduler). Declaration: Preemptive_Scheduler : **aadlboolean applies to** (processor); Property Association Example: *Preemptive_Scheduler => false;*

Table A-4: *Deployment Properties (continued)*

Property	Description
Thread_Limit	A property that specifies the maximum number of threads supported by a processor or virtual processor. It can be specified only for processors and virtual processors, and is of type **aadlinteger**. The upper limit for its value is *Max_Thread_limit*. Declaration: Thread_Limit: **aadlinteger** 0 .. Max_Thread_Limit **applies to** (processor, virtual processor); Property Association Example: *Thread_Limit=> 15;*
Priority_Map	A property that specifies a mapping of AADL priorities into priorities of an underlying real-time operating system. This map consists of a list of one or more entries such that each entry maps one AADL priority to one underlying real-time operating system priority. An entry is of type *Priority_Mapping*, which is a **record** type that consists of a pair of **aadlinteger** values: *Aadl_Priority* and *RTOS_Priority*. Declaration: Priority_Map: **list of** Priority_Mapping **applies to** (processor); Property Association Examples: *Priority_Map =>((Aadl_Priority => 1, RTOS_Priority => 2), (Aadl_Priority => 2, RTOs_Priority => 3));* *Priority_Map => (Aadl_Priority => 4, RTOS_Priority => 5);*
Priority_Range	A property that specifies the range of thread priority values acceptable to a processor or virtual processors. It can be specified only for processors and virtual processors. The property type is a **range of aadlinteger**. Declaration: Priority_Range: **range of aadlinteger applies to** (processor, virtual processors); Property Association Example: *Priority_Range => 1 .. 2;*

A.5.2 Thread-Related Properties

Properties useful in describing thread and thread execution character-istics are listed in Table A-5. The table includes property declarations taken from the standard and examples of property associations. Since all standard and user-defined properties that can be specified for at least one component of any category can be specified for an abstract component, the abstract component category is not included in the table.

Table A-5: *Thread-Related Properties*

Property	Description
Dispatch_Protocol	A property that specifies the dispatch behavior of a thread. It can be specified for a thread, device, or virtual processor. This is a property of type *Supported_Dispatch_Protocols* that is defined in the *AADL_project* property set. A single value *Periodic* is declared in the Standard for the type *Supported_Dispatch_Protocols*. Other values are designated as *project specified* and can be defined by a user. Some example values that a user may choose to add include *Periodic, Sporadic, Aperiodic, Timed, Hybrid,* and, *Background*. Declaration: Dispatch_Protocol: *Supported_Dispatch_Protocols* **applies to** (thread, device, virtual processor); Property Association Examples (assumes *Sporadic* and *Aperiodic* have been defined in *AADL_project*): *Dispatch_Protocol = Aperiodic ;* *Dispatch_Protocol => Sporadic ;*
Dispatch_Trigger	A property that specifies a list of one or more ports that can trigger the dispatch of a thread or device. It can be specified for threads and devices. Declaration: Dispatch_Trigger: **list of reference** (port) **applies to** (device, thread); Property Association Example: *Dispatch_Trigger => (reference(engage1));*

Table A-5: *Thread-Related Properties (continued)*

Property	Description
POSIX_Scheduling_Policy	A property specifies the scheduling policy assigned to a given thread using the scheduling protocols defined by the POSIX 1003.1b standard. The property is an **enumeration** with the values of *SCHED_FIFO, SCHED_RR or SCHED_OTHER*. In a POSIX 1003.1b architecture, the policy allows the scheduler to choose the thread to run when several threads have the same fixed priority. If a thread does not define the *POSIX_Scheduling_Policy* property, it has the value *SCHED_FIFO*. The policy semantics are • *SCHED_FIFO* implements a FIFO scheduling protocol on a set of equal fixed priority : a thread stays on the processor until it has terminated or until a highest priority thread is released. • *SCHED_RR,* similar to SCHED_FIFO except that a quantum is used. At the end of the quantum, the running thread is pre-empted from the processor and an equal priority thread has to be released. • *SCHED_OTHER* is defined by POSIX policy implementers; usually implements a timing sharing scheduling protocol. Declaration: POSIX_Scheduling_Policy : **enumeration** (SCHED_FIFO, SCHED_RR, SCHED_OTHERS) **applies to** (thread, thread group); Property Association Example: *POSIX_Scheduling_Policy => SCHED_FIFO;*
Priority	An inherited property that specifies the priority of a thread. It can be specified for threads, thread groups, processes, systems, and devices. It is a property of type **aadlinteger** whose value is within the range supported by the processor to which the thread is bound. If specified for a device, it applies to the device driver. Declaration: Priority: **inherit aadlinteger applies to** (thread, thread group, process, system, device); Property Association Example: *Priority => 3;*

continues

Table A-5: *Thread-Related Properties (continued)*

Property	Description
Criticality	A property that specifies the criticality level of a thread. It can be specified for threads and thread groups. It is a property of type **aadlinteger**. This property is used by maximum urgency first scheduling protocols and can be used by project specific scheduling protocols. Declaration: Criticality: **aadlinteger applies to** (thread, thread group); Property Association Example: *Criticality => 2;*
Time_Slot	A property that specifies a list of one or more statically allocated slots on a timeline. It can be specified for threads, thread groups, processes, virtual processors, and systems. It is a property of type **list of aadlinteger**. Scheduling protocols with a time slot allocation approach such as the protocol for scheduling partitions on a static timeline use this property. Declaration: Time_Slot: **list of aadlinteger applies to** (thread, thread group, process, virtual processor, system); Property Association Example: *Time_Slot => (5);* *Time_Slot = > (4, 5) ;*
Concurrency_Control_ Protocol	A property that specifies the concurrency control protocol used to ensure mutually exclusive access such as restricting access to a critical region or shared data. It can be specified for data components. It is of type *Supported_Concurrency_Control_Protocols* that is defined in the *AADL_project* property set. *Supported_Concurrency_ Control_Protocols* is an **enumeration** property type with a single standard value (*None_Specified*). Other values are designated as *project-specified* and can be defined by a user. If no value is specified the value is *None_Specified* (i.e., no concurrency control protocol). Declaration: Concurrency_Control_Protocol: Supported_Concurrency_Control_ Protocols **applies to** (data); Property Association Example: *Concurrency_Control_Protocol = > Spin_Lock;*

Table A-5: *Thread-Related Properties (continued)*

Property	Description
Urgency	A property that specifies the urgency with which an event at an in port is to be serviced relative to other events arriving at or queued at other in ports of the same thread. It can be specified for ports and subprograms. It is of property type **aadlinteger** whose maximum value is *Max_Urgency*. *Max_Urgency* is a property constant declared in the *AADL_project* property set. A user specifies its value such that the larger the value the higher is the urgency. Declaration: Urgency: **aadlinteger** 0 .. Max_Urgency **applies to** (port, subprogram); Property Association Example: *Urgency* => 1;
Dispatch_Able	A property that specifies whether a thread should be dispatched in a given mode. This allows you to specify that in a specific mode a thread is idle without the thread getting dispatched and immediately completing its execution. Declaration: Dispatch_Able: **aadlboolean applies to** (thread); Property Association Example: *Dispatch_Able* => *false **in modes** (idle);*
Dequeue_Protocol	A property that specifies dequeuing options. It can be specified for event ports and event data ports. It is a property of type **enumeration** with the values *OneItem, MultipleItems,* and *AllItems*. The semantics of the values are • *OneItem:* A single item is dequeued at input time and made available to the source text unless the queue is empty. The *Next_Value* service call has no effect. • *AllItems:* All items that are frozen at input time are dequeued and made available to the source text via the port variable, unless the queue is empty. Individual items become accessible as port variable values through the *Next_Value* service call. Any element in the frozen queue that is not retrieved through the *Next_Value* service call is discarded (i.e., is removed from the queue and is not available at the next input time).

continues

Table A-5: *Thread-Related Properties (continued)*

Property	Description
Dequeue_Protocol (continued)	• *MultipleItems:* Multiple items can be dequeued, one at a time from the frozen queue and made available to the source text via the port variable. One item is dequeued and its value made available via the port variable with each *Next_Value* service call. Any items not dequeued remain in the queue and are available at the next input time. If the *Dequeued_Items* property is set, it imposes a maximum on the number of elements that are made accessible to a thread at input time when the *Dequeue_Protocol* property is set to *AllItems* or *MultipleItems*. The default property value is OneItem. The *Next_Value* service call is defined in Table A-12. Declaration: Dequeue_Protocol: **enumeration** (OneItem, MultipleItems, AllItems) => OneItem **applies to** (event port, event data port); Property Association Example: *Dequeue _Protocol => OneItem;*
Dequeued_Items	A property that specifies the maximum number of items available to an application via an event or event data port at input time. It can be specified for event ports and event data ports. It is a property of type **aadlinteger** whose value cannot exceed the value of the *Queue_Size* property for the port. Declaration: Dequeued_Items: **aadlinteger applies to** (event port, event data port); Property Association Example: *Dequeued_Items => 5;* *Dequeued_Items = > 10* **applies to** *monitor.error_port;*
Mode_Transition_ Response	A property that specifies whether a mode transition occurs immediately due to an emergency situation or whether it is planned such that the completion of thread execution can be coordinated before performing the mode transition. It can be specified only for mode transitions. It is a property of type **enumeration** with the values *emergency* and *planned*. If a value is not specified, the value is planned.

Table A-5: *Thread-Related Properties (continued)*

Property	Description
Mode_Transition_ Response (continued)	Declaration: Mode_Transition_Response: **enumeration** (emergency, planned) **applies to** (mode transition); Property Association Example: *Mode_Transition_Response => emergency;*
Resumption_Policy	A property that specifies whether, as result of a mode transition activation, a component with modes starts in the initial mode or resumes in the mode at the time of its deactivation. It can be specified for threads, thread groups, processes, systems, devices, processors, memory, buses, systems, virtual processors, virtual buses, and subprograms. It is a property of type enumeration with the values *restart* and *resume*. Declaration: Resumption_Policy: **enumeration** (restart, resume) **applies to** (thread, thread group, process, system, device, processor, memory, bus, system, virtual processor, virtual bus, subprogram); Property Association Example: *Resumption_Policy => restart;*
Active_Thread_ Handling_Protocol	An inherited property that specifies the protocol for handling execution at the instant of an actual mode switch. It can be specified for threads, thread groups, processes, and systems and is of property type *Supported_Active_Thread_Handling_Protocols*. Supported_Active_Thread_Handling_Protocols is declared in the *AADL_ Project* property set where a single value [*stop/abort*] is defined. Other values are designated as *project-specified* and can be defined by a user. Some values that may be added include *suspend, complete_one,* or *complete_all.* Declaration: Active_Thread_Handling_Protocol: **inherit** Supported_Active_ Thread_Handling_Protocols => abort **applies to** (thread, thread group, process, system); Property Association Examples (assumes *complete_one* and *complete_ all* have been defined in *AADL_project*): *Active_Thread_Handling_Protocol: => complete_one ;* *Active_Thread_Handling_Protocol: = > complete_all ;*

continues

Table A-5: *Thread-Related Properties (continued)*

Property	Description
Active_Thread_Queue_ Handling_Protocol	An inherited property that specifies the protocol for handling the content of any event port or event data port queue of a thread at the instant of a mode switch. It can be specified for threads, thread groups, processes, and systems. It is a property of type **enumeration** with values *flush* and *hold*. *Flush* empties the queue. *Hold* keeps the content in the queue of the thread being deactivated until it is reactivated. If no value is specified the value is *flush*. Declaration: Active_Thread_Queue_Handling_Protocol: **inherit enumeration** (flush, hold) => flush **applies to** (thread, thread group, process, system); Property Association Example: Active_Thread_Queue_Handling_Protocol => *flush;* Mode_Transition_Response = > *hold* **applies to** *controller_system;*
Deactivation_Policy	A property that specifies whether a process is to be unloaded when it is deactivated. It can be specified for threads, processes, virtual processors, and processors. It is a property of type **enumeration** with the values *inactive* and *unload*. If the policy is unload, the process is unloaded on deactivate and loaded on activate. If no value is assigned to the property, the value is *inactive* and the process is loaded during startup and is not unloaded when deactivated. Declaration: Deactivation_Policy: **enumeration** (inactive, unload) => inactive **applies to** (thread, process, virtual processor, processor); Property Association Example: *Deactivation_Policy => inactive;*
Runtime_Protection	An inherited property that specifies whether a process requires runtime enforcement of address space protection. It can be specified for processes, systems, or virtual processors. It is a property of type **addl_boolean**. If no value is specified the value is **true**. Declaration: Runtime_Protection : **inherit aadlboolean applies to** (process, system, virtual processor); Property Association Example: *Runtime_Protection =>* **false***;*

Table A-5: *Thread-Related Properties (continued)*

Property	Description
Subprogram_Call_Type	A property that specifies whether a call is to be performed as synchronous or semi-synchronous. It can be specified only for subprograms. It is a property of type **enumeration** with the values *Synchronous* and *SemiSynchronous*. For a semi-synchronous call, the caller continues execution and waits for the result at a later time (*Await_result* service call). If no property value is specified, the value is *Synchronous*. Declaration: Subprogram_Call_Type: **enumeration** (Synchronous, SemiSynchronous) => Synchronous **applies to** (subprogram); Property Association Example: *Subprogram_Call_Type => Synchronous;*
Synchronized_ Component	An inherited property that specifies whether a periodic thread will be synchronized with transitions into and out of a mode. It can be specified for threads, thread groups, processes, and systems. It is a property type of **aadlboolean**. If the value is true, the thread will be synchronized at the hyperperiod. If no value is specified for the property, the value is **true**. Declaration: Synchronized_Component: **inherit aadlboolean** => **true applies to** (thread, thread group, process, system); Property Association Example: *Synchronized_Component => **false**;*

A.5.3 Timing Properties

Properties useful in specifying timing related aspects are listed in Table A-6. The table includes the property declaration taken from the standard and examples of property associations involving each property. Since all standard and user-defined properties that can be specified for at least one component of any category can be specified for an abstract component, the abstract component category is not included in the table.

Table A-6: *Timing Properties*

Property	Description
Activate_Deadline	A property that specifies the maximum amount of time allowed for the execution of a thread's activation sequence. It can be specified only for threads. The property type is *Time* and the value must be positive. The standard *Time* units are ps (picoseconds), ns (nanoseconds), us (microseconds), ms (milliseconds), sec (seconds), min (minutes), and hr (hours). Declaration: Activate_Deadline: Time **applies to** (thread); Property Association Example: *Activate_Deadline => 5 us;*
Activate_Execution_Time	A property that specifies the minimum and maximum execution time that a thread will use to execute its activation sequence (e.g., when a thread becomes active as part of a mode switch). It can be specified only for threads. The specified execution times assume no runtime errors in the execution and include the time required to execute any service calls executed by the thread, but excludes any time spent by another thread executing remote procedure calls in response to a remote subprogram call made by the thread. The property type is *Time_Range* such that the range specifies the minimum and maximum time for execution. These times are expressed using the standard *Time* units of ps (picoseconds), ns (nanoseconds), us (microseconds), ms (milliseconds), sec (seconds), min (minutes), and hr (hours). Declaration: Activate_Execution_Time: Time_Range **applies to** (thread); Property Association Example: *Activate_Execution_Time => 5 us.. 10 us;*
Compute_Deadline	A property that specifies the maximum amount of time allowed for the execution of a thread's compute sequence. It can be specified for threads, devices, subprograms, subprogram accesses, event ports, and event data ports. When specified for a subprogram, event port, or event data port, the value applies to the associated thread that is dispatched by an initiating call, event, or event data (message) occurrence. When specified for a subprogram access feature, it applies to the thread executing the remote procedure call in response to the remote subprogram call.

Table A-6: *Timing Properties (continued)*

Property	Description
Compute_Deadline (continued)	The values specified for this property for a thread are bounds on the values specified for features such that the value specified for a feature must not exceed the value of its associated thread. The value of this property is constrained by the values of *the* properties *Deadline* and *Recover_Deadline* such that *Compute_Deadline* + *Recover_Deadline* ≤ *Deadline*. The value is of type *Time* and must be positive. The standard units are ps (picoseconds), ns (nanoseconds), us (microseconds), ms (milliseconds), sec (seconds), min (minutes), and hr (hours). Declaration: Compute_Deadline: Time **applies to** (thread, device, subprogram, subprogram access, event port, event data port); Property Association Example: *Compute_Deadline => 10 ms;*
Compute_Execution_Time	A property that specifies the amount of time that a thread will execute after it has been dispatched, before it begins waiting for another dispatch. It can be specified for threads, devices, subprograms, event ports, and event data ports. When specified for a subprogram, event port, or event data port, the value applies to the associated thread that is dispatched by an initiating call, event, or event data (message) occurrence. When specified for a subprogram (access) feature, it applies to the thread executing the remote procedure call in response to a remote subprogram call. The value specified for a feature must not exceed the *Compute_Execution_Time* of its associated thread. It is of type *Time_Range* that specifies a minimum and maximum execution time in the absence of runtime errors. The specified execution time includes the time required to execute any service calls that are executed by the thread, but excludes any time spent by another thread executing remote procedure calls in response to a remote subprogram call made by the thread. The values specified for this property for a thread are bounds on the values specified for the thread's associated features. Declaration: Compute_Execution_Time: Time_Range **applies to** (thread, device, subprogram, event port, event data port); Property Association Example: *Compute_Execution_Time => 2 ms .. 5 ms;*

continues

Table A-6: *Timing Properties (continued)*

Property	Description
Client_Subprogram_ Execution_Time	A property that specifies the length of time it takes to execute the client portion of a remote subprogram call. It can be specified only for subprograms. The property type is *Time_Range*. The values must be positive and are expressed using the standard *Time* units of ps (picoseconds), ns (nanoseconds), us (microseconds), ms (milliseconds), sec (seconds), min (minutes), and hr (hours). Declaration: Client_Subprogram_Execution_Time: Time_Range **applies to** (subprogram); Property Association Example: *Client_Subprogram_Execution_Time => 2 ms .. 5 ms;*
Deactivate_Deadline	A property that specifies the maximum amount of time allowed for the execution of a thread's deactivation sequence. It can be specified only for threads. The property type is *Time* that has the standard units of ps (picoseconds), ns (nanoseconds), us (microseconds), ms (milliseconds), sec (seconds), min (minutes), and hr (hours). The value must be positive. Declaration: Deactivate_Deadline: Time **applies to** (thread); Property Association Example: *Deactivate _Deadline => 2 us;*
Deactivate_Execution_ Time	A property specifies the amount of time that a thread will execute its deactivation sequence when the thread is deactivated as part of a mode switch. It can be specified only for threads. The range expression specifies a minimum and maximum execution time in the absence of runtime errors. The time includes the time required to execute any service calls executed by a thread, but excludes any time spent by another thread executing remote procedure calls in response to a remote subprogram call made by the thread. It is of type *Time_Range*. The minimum and maximum values must be positive and are of type *Time* that has the standard units of ps (picoseconds), ns (nanoseconds), us (microseconds), ms (milliseconds), sec (seconds), min (minutes), and hr (hours).

Table A-6: *Timing Properties (continued)*

Property	Description
Deactivate_Execution_ Time (continued)	Declaration: Deactivate_Execution_Time: Time_Range **applies to** (thread); Property Association Example: *Deactivate_Execution_Time => 2 ms .. 5 ms;*
Deadline	An inherited property that specifies the maximum amount of time allowed between a thread dispatch and the time that thread begins waiting for another dispatch. It can be specified for threads, thread groups, processes, systems, devices, and virtual processors; is of type *Time*; and must be a positive value. This property places a limit on the properties *Compute_Deadline* and *Recover_Deadline* such that *Compute_Deadline* + *Recover_Deadline* ≤ *Deadline*. A *Deadline* property may not be specified for threads that are dispatched as background threads. Declaration: Deadline: **inherit** Time => Period **applies to** (thread, thread group, process, system, device, virtual processor) Property Association Example: *Deadline => 20 ms;*
First_Dispatch_Time	An inherited property that specifies the time of the first dispatch request and applies to threads and thread groups. The property type is *Time* and must be a positive value. Declaration: First_Dispatch_Time **: inherit** Time **applies to** (thread, thread group); Property Association Example: *First_Dispatch_Time => 5ms;*
Dispatch_Jitter	An inherited property that specifies a maximum bound on the lateness of a thread's dispatch. It can be specified for threads and thread groups. A periodic thread is dispatched based upon a fixed delay (i.e., its period). It may occur in a system that the thread's dispatch event is delayed. This delay can be specified with the *Dispatch_Jitter* property for any thread that is dispatched multiple times (e.g., Periodic, Sporadic). The property type is *Time* and must be a positive value.

continues

Table A-6: *Timing Properties (continued)*

Property	Description
Dispatch_Jitter *(continued)*	Declaration: Dispatch_Jitter: **inherit** Time **applies to** (thread, thread group); Property Association Example: *Dispatch_Jitter => 2 ms;*
Dispatch_Offset	An inherited property that specifies a dispatch time offset for a thread. It applies only to periodic threads. The offset indicates the amount of time by which the dispatch of a thread is offset relative to its period. The property type is *Time* and must be a positive value. Declaration: Dispatch_Offset: **inherit** Time **applies to** (thread); Property Association Example: *Dispatch_Offset => 2 ms;*
Execution_Time:	A property that specifies the amount of execution time a virtual processor can allocate to threads or virtual processors it schedules. It can be specified only for virtual processors. The property type is *Time* and must be a positive value. Declaration: Execution_Time: Time **applies to** (virtual processor); Property Association Example: *Execution_Time => 20 ms;*
Finalize_Deadline	A property that specifies the maximum time allowed for the execution of a thread's finalization sequence. It can be specified only for threads; is a property of type *Time*; and must be a positive value. The standard units are ps (picoseconds), ns (nanoseconds), us (microseconds), ms (milliseconds), sec (seconds), min (minutes), and hr (hours). Declaration: Finalize_Deadline: Time **applies to** (thread); Property Association Example: *Finalize_Deadline => 2 us;*

Table A-6: *Timing Properties (continued)*

Property	Description
Finalize_Execution_Time: Time	A property that specifies the amount of time that a thread will execute its finalization sequence. It can be specified only for threads and is a property of type *Time_Range* that establishes a minimum and maximum execution time in the absence of runtime errors. The specified execution time includes the time required to execute any service calls executed by the thread, but excludes any time spent by another thread executing remote procedure calls in response to a remote subprogram call made by the thread. Declaration: Finalize_Execution_Time: Time_Range **applies to** (thread); Property Association Example: *Finalize_Execution_Time => 1 us .. 10 us;*
Initialize_Deadline	A property that specifies the maximum amount of time allowed between the time a thread executes its initialization sequence and the time that thread begins waiting for a dispatch. It can be specified only for threads. The property type is *Time* and must be a positive value. The standard units are ps (picoseconds), ns (nanoseconds), us (microseconds), ms (milliseconds), sec (seconds), min (minutes), and hr (hours). Declaration: Initialize_Deadline: Time **applies to** (thread); Property Association Example: *Initialize_Deadline => 200 us;*
Initialize_Execution_Time	A property that specifies the amount of time that a thread will execute its initialization sequence. It applies to threads and is a property of type *Time_Range* that establishes a minimum and maximum execution time in the absence of runtime errors. The specified execution time includes all time required to execute any service calls executed by the thread, but excludes any time spent by another thread executing remote procedure calls in response to a remote subprogram call made by the thread. Declaration: Initialize_Execution_Time: Time_Range **applies to** (thread); Property Association Example: *Initialize _Execution_Time => 2 ms .. 3 ms;*

continues

Table A-6: *Timing Properties (continued)*

Property	Description
Load_Deadline	A property that specifies the maximum amount of elapsed time allowed between the time a process begins and completes loading. It can be specified for processes and systems. Its property type is *Time* and must be a positive value. The standard units are ns (nanoseconds), us (microseconds), ms (milliseconds), sec (seconds), min (minutes), and hr (hours). Declaration: Load_Deadline: Time **applies to** (process, system); Property Association Example: *Load _Deadline => 1 ms;*
Load_Time	A property that specifies the amount of execution time required to load the binary image associated with a process. It can be specified for processes and systems and is a property of type *Time_Range* that establishes a minimum and maximum load time in the absence of runtime errors. When applied to a system, the property specifies the amount of time it takes to load the binary image of data components declared within the system implementation and shared across processes and their address spaces. Declaration: Load_Time: Time_Range **applies to** (process, system); Property Association Example: *Load_Time => 2 ms .. 3 ms;*
Period	An inherited property that specifies the time interval between successive dispatches of a thread whose scheduling protocol is *periodic*, or the minimum interval between successive dispatches of a thread whose scheduling protocol is *sporadic*. It can be specified for threads, thread groups, processes, systems, devices, and virtual processors. The property type is *Time* and must be a positive value. The standard units are ns (nanoseconds), us (microseconds), ms (milliseconds), sec (seconds), min (minutes), and hr (hours). Declaration: Period: **inherit** Time **applies to** (thread, thread group, process, system, device, virtual processor); Property Association Example: *Period => 20 ms;*

Table A-6: *Timing Properties (continued)*

Property	Description
Recover_Deadline	A property that specifies the maximum amount of time allowed between the time when a detected error occurs and the time a thread begins waiting for another dispatch. It can be specified only for threads. It is a property of type *Time* and must be a positive value. The value must not be greater than that specified for the Period of the thread and is constrained by the values of the properties *Compute_Deadline and Recover_Deadline* such that *Compute_Deadline + Recover_Deadline ≤ Deadline*. It is not legal to specify a value for this property for threads whose dispatch protocol is background. Declaration: Recover_Deadline: Time **applies to** (thread); Property Association Example: *Recover_Deadline => 10 ms;*
Recover_Execution_Time	A property that specifies the amount of time a thread will execute after an error has occurred, before it begins waiting for another dispatch. It can be specified only for threads. It is a property of type *Time_Range* that establishes a minimum and maximum execution time in the absence of runtime errors. The specified execution time includes the time required to execute any service calls executed by the thread, but excludes any time spent by another thread executing remote procedure calls in response to a remote subprogram call made by the thread. Declaration: Recover_Execution_Time: Time_Range **applies to** (thread); Property Association Example: *Recover_Execution_Time => 2 ms .. 3 ms;*
Startup_Deadline	A property that specifies the deadline for processor, virtual processor, process, and system initialization. It can be specified for processors, virtual processors, processes, and systems. The property type is *Time* and must be a positive value. The standard units are ps (picoseconds), ns (nanoseconds), us (microseconds), ms (milliseconds), sec (seconds), min (minutes), and hr (hours).

continues

Table A-6: *Timing Properties (continued)*

Property	Description
Startup_Deadline *(continued)*	Declaration: Startup_Deadline: Time **applies to** (processor, virtual processor, process, system); Property Association Example: *Startup_Deadline => 5 ms;*
Startup_Execution_Time	A property that specifies the execution time for the initialization of a virtual processor or process. It can be specified for virtual processors, processors, processes, and systems. It is a property of type *Time_Range* that establishes a minimum and maximum execution time in the absence of runtime errors. These values are of type *Time* and must be positive. Declaration: Startup_Execution_Time: Time_Range **applies to** (virtual processor, processor, process, system); Property Association Example: *Startup_Execution_Time => 1 ms .. 2 ms;*
Clock_Jitter	A property that specifies a time unit value that gives the maximum time between the start of clock interrupt handling on any two processors in a multiprocessor system. It can be specified for processors and systems. The property type is *Time* and must be a positive value. The standard units are ns (nanoseconds), us (microseconds), ms (milliseconds), sec (seconds), min (minutes), and hr (hours). Declaration: Clock_Jitter: Time **applies to** (processor, system); Property Association Example: *Clock_Jitter => 10 us;*
Clock_Period	A property that specifies the time interval between two clock interrupts. It can be specified for processors and systems. The property type is *Time* and must be a positive value. The standard units are ns (nanoseconds), us (microseconds), ms (milliseconds), sec (seconds), min (minutes), and hr (hours). The numeric value must be a positive number.

Table A-6: *Timing Properties (continued)*

Property	Description
Clock_Period (continued)	Declaration: Clock_Period: Time **applies to** (processor, system); Property Association Example: *Clock_Period => 10 us;*
Clock_Period_Range	A property that specifies a time range that represents the minimum and maximum values assignable to the *Clock_Period* property. It can be specified for processors and systems. The property type is *Time_Range*. The minimum and maximum values must be positive and are expressed using the standard *Time* units of ps (picoseconds), ns (nanoseconds), us (microseconds), ms (milliseconds), sec (seconds), min (minutes), and hr (hours). Declaration: Clock_Period_Range: Time_Range **applies to** (processor, system); Property Association Example: *Clock_Period_Range => 1us .. 3 ms;*
Process_Swap_ Execution_Time	A property specifies the amount of execution time necessary to perform a context swap between two threads contained in different processes that the property applies to processors. It is a property of type *Time_Range* that establishes a minimum and maximum swap time in the absence of runtime errors. The minimum and maximum values must be positive and are expressed using the standard *Time* units of ps (picoseconds), ns (nanoseconds), us (microseconds), ms (milliseconds), sec (seconds), min (minutes), and hr (hours). Declaration: Process_Swap_Execution_Time: Time_Range **applies to** (processor); Property Association Example: *Process_Swap_Execution_Time => 10 us .. 20 us;*
Reference_Processor	An inherited property that defines the processor that is the basis for specifying execution times. It can be specified for subprograms, subprogram groups, threads, thread groups, processes, devices, and systems. It is a property of type **classifier**. When code is bound to another processor type, a *Scaling_Factor* associated with the other processor is used to determine the execution times relative to the reference processor, unless a binding specific execution time value has been defined.

continues

Table A-6: *Timing Properties (continued)*

Property	Description
Reference_Processor (continued)	Declaration: Reference_Processor: **inherit classifier** (processor) **applies to** (subprogram, subprogram group, thread, thread group, process, device, system); Property Association Example: *Reference_Processor => **classifier**(RT_3GHz.basic);*
Scaling_Factor	An inherited property that specifies the relative speed (speed factor) of a processor with respect to a reference processor. It can be specified for processors and systems and is of type **aadlreal**. Declaration: Scaling_Factor : **inherit aadlreal applies to** (processor, system); Property Association Example: *Scaling_Factor => 0.5;*
Scheduler_Quantum	An inherited property that specifies the maximum time a thread can hold a processor without being preempted. It can be specified only for processors. A quantum can be used with any user-defined scheduler. It is typically used in time sharing scheduling and in POSIX 1003.1b scheduling (with the SCHED_RR policy). If the quantum is not specified for a given processor, the quantum has a positive infinitesimal value. It is a property of type *Time* and must be a positive value. Declaration: Scheduler_Quantum : **inherit** Time **applies to** (processor); Property Association Example: *Scheduler_Quantum => 300 us;*
Thread_Swap_ Execution_Time	A property that specifies the amount of execution time necessary for performing a context swap between two threads contained within the same process. It can be specified for processors and systems and is a property of type *Time_Range* that establishes a minimum and maximum swap time in the absence of runtime errors. The minimum and maximum values must be positive and are expressed using the standard *Time* units of ps (picoseconds), ns (nanoseconds), us (microseconds), ms (milliseconds), sec (seconds), min (minutes), and hr (hours).

Table A-6: *Timing Properties (continued)*

Property	Description
Thread_Swap_ Execution_Time (continued)	Declaration: Thread_Swap_Execution_Time: Time_Range **applies to** (processor, system); Property Association Example: *Thread_Swap_Execution_Time => 2 us .. 3 us;*
Frame_Period	A property that specifies the time period of a major frame in a static scheduling protocol, such as a cyclic executive. It can be specified for processors and virtual processors. The property type is *Time* and must be a positive value. The standard units are ns (nanoseconds), us (microseconds), ms (milliseconds), sec (seconds), min (minutes), and hr (hours). Declaration: Frame_Period: Time **applies to** (processor, virtual processor); Property Association Example: *Frame_Period => 100 ms;*
Slot_Time	A property that specifies the time period of a slot in major frame in a static scheduling protocol, such as a cyclic executive, if the protocol uses fixed slot times. It can be specified for processors and virtual processors. The property type is *Time* and must be a positive value. The standard units are ns (nanoseconds), us (microseconds), ms (milliseconds), sec (seconds), min (minutes), and hr (hours). Declaration: Slot_Time: Time **applies to** (processor, virtual processor); Property Association Example: *Slot_Time => 10 ms;*

A.5.4 Communication Properties

Properties useful in describing communication, including connection topology and queuing characteristics are listed in Table A-7. The table includes property declarations taken from the standard and examples of property associations. Since all standard and user-defined properties that can be specified for at least one component of any category can be specified for an abstract component, the abstract component category is not included in the table.

Table A-7: *Communication-Related Properties*

Property	Description
Fan_Out_Policy	A property that specifies how the output of a port with multiple outgoing connections is distributed. The property can only be specified for a port. It is a property of type **enumeration** with values of *Broadcast, RoundRobin, Selective,* and *OnDemand. Broadcast* sends to all recipients, *RoundRobin* sends to one recipient at a time in order, *Selective* sends to one recipient based on data content, and *OnDemand* sends to the next recipient waiting on a port for dispatch. *Broadcast, RoundRobin,* and *Selective* pass on data and events without queuing. *OnDemand* requires a queue that is serviced by the recipients. Declaration: Fan_Out_Policy: **enumeration** (Broadcast, RoundRobin, Selective, OnDemand) **applies to** (port); Property Association Example: Fan_Out_Policy => *Broadcast;*
Connection_Pattern	A property that specifies the individual connections between arrays of ports. The property can be specified only for a connection. The property is of type *Supported_Connection_Patterns* that is an **enumeration** type with values *One_To_One, All_To_All, One_To_All, All_To_One, Next, Previous, Cyclic_Next,* and *Cyclic_Previous.* If the property is not specified the value is *One_to_One.* Declaration: Connection_Pattern: **list of** Supported_Connection_Patterns **applies to** (connection); Property Association Example: Connection_Pattern => *((One_To_All)) ;*

Table A-7: *Communication-Related Properties (continued)*

Property	Description
Connection_Set	A property that specifies a list of specific source elements and destination elements of a semantic connection, identified by their array indices. It can be specified only for a connection. It is of type *Connection_Pair* that is a **record** type with two fields, *src* and *dst*. Each of these fields is a list of integers. Declaration: Connection_Set: **list of** Connection_Pair **applies to** (connection); Property Association Examples: *Connection_Set => [src = > (1,3, 5); dst=> (5, 3,1);];* *Connection_Set => [(src = > (1,3, 5); dst=> (5, 3,1);),(src = > (2,4); dst=> (6));] ;*
Overflow_Handling_ Protocol	A property that specifies the runtime behavior of a thread when an event arrives and the queue is full. It can be specified for event ports, event data ports, and subprogram accesses. It is of type enumeration with the values *DropOldest, DropNewest,* and *Error.* *DropOldest* removes the oldest event from the queue and adds the new arrival. *DropNewest* ignores the newly arrived event. *Error* causes the thread's error recovery to be invoked. If no value is specified the value is *DropOldest.* Declaration: Overflow_Handling_Protocol: **enumeration** (DropOldest, Drop-Newest, Error) => DropOldest **applies to** (event port, event data port, subprogram access); Property Association Example: *Overflow_Handling_Protocol => DropOldest;*
Queue_Processing_ Protocol	A property that specifies the protocol for processing elements in a queue. It can be specified for event ports, event data ports, and subprogram accesses. It is a property of type that is an **enumeration** type with the standard value *FIFO.* Other values are *project-specified* that can be defined by a user. If no value is specified for the property, the value is *FIFO.* Declaration: Queue_Processing_Protocol: Supported_Queue_Processing_Protocols => FIFO **applies to** (event port, event data port, subprogram access);

continues

Table A-7: *Communication-Related Properties (continued)*

Property	Description
Queue_Processing_ Protocol (continued)	Property Association Example (assumes LIFO has been defined by a user): *Queue_Processing_Protocol => LIFO;*
Queue_Size	A property that specifies the size of a queue for an event, event data port, or a subprogram access feature of a data component being shared via data access. It can be specified for event ports, event data ports, and subprogram accesses. It is a property of type **aadlinteger**. The maximum value that can be assigned to the property is the value of the *Max_Queue_Size*. When specified for a subprogram access it represents the queue for remote subprogram calls. When specified for a data component it represents the queue used in resource locking. Declaration: Queue_Size: **aadlinteger** 0 .. Max_Queue_Size => 1 **applies to** (event port, event data port, subprogram access); Property Association Example: *Queue_Size => 10;*
Required_Connection	A property that specifies whether a port or subprogram requires a connection. It can be specified for any feature. It is a property of type **aadlboolean**. If the value of this property is **false**, it is assumed that the component can function without this port or subprogram access feature being connected. If a value is not specified for the property, the value is **true** (a connection is required). Declaration: Required_Connection : **aadlboolean => true applies to** (feature); Property Association Example: *Required_Connection => false;*
Timing	A property that specifies the timing of port connections. It can be specified only for port connections. It is a property of type **enumeration** with the values *sampled, immediate,* and *delayed*. If no value is assigned to the property the value is *sampled* (i.e., the receiving component samples at dispatch or during execution).

Table A-7: *Communication-Related Properties (continued)*

Property	Description
Timing *(continued)*	Declaration: Timing : **enumeration** (sampled, immediate, delayed) **=>** sampled **applies to** (port connection); Property Association Example: *Timing => immediate;*
Transmission_Type	A property that specifies whether the transmission across a data port connection is initiated by the sender (push) or by the receiver (pull). It can be specified for data ports, port connections, buses, and virtual buses. It is a property of type **enumeration** with the values *push* and *pull*. If no value is assigned to the property the value is *push* (i.e., the transmission is initiated by the sender). A *pull* transmission type results in data being transmitted at the rate of the receiver. For event data ports or event ports, a *pull* transmission type results in events or event data queued with the sender to be transmitted upon receiver request. When associated with a connection, the property represents the transmission type for the connection. When associated with a port, the property represents the transmission type of the port. When associated with a bus or virtual bus, the property represents the transmission type that is provided by the bus or protocol. Declaration: Transmission_Type: **enumeration** (push, pull) **applies to** (data port, port connection, bus, virtual bus); Property Association Example: *Transmission_Type => pull;*
Input_Rate	A property that specifies the number of inputs per dispatch or per second of data, events, event data, or subprogram calls. It can be specified only for ports and is a property of type *Rate_Spec*. *Rate_Spec* is a record type consisting of three fields: *Value_Range, Rate_Unit*, and *Rate_Distribution* where *Value_Range* is a range of real number; *Rate_Unit* is either *PerSecond* or *PerDispatch*; and *Rate_Distribution* can be assigned the **enumeration** values of *Fixed* or values specified by a user (i.e., *project-specified*). If a rate and unit value are not specified, the values are (1.0..1.0) input per thread dispatch. If no rate distribution is specified, the value is *Fixed*.

continues

Table A-7: *Communication-Related Properties (continued)*

Property	Description
Input_Rate *(continued)*	Declaration: Input_Rate: Rate_Spec => [Value_Range => 1.0 .. 1.0; Rate_Unit => PerDispatch; Rate_Distribution => Fixed;] **applies to** (port); Property Association Example: *Input_Rate => [Value_Range => 50.0 .. 100.0; Rate_Unit => Per-Second; Rate_Distribution => Fixed;];*
Input_Time	A property that specifies the amount of execution time following a dispatch before the input is frozen on a given port. The Input time can be expressed relative to execution time at start or completion of execution, or relative to the start time. It can be specified only for a port and is a list of *IO_Time_Spec* values. *IO_Time_Spec* is a record type with two fields *Offset* and *Time*. *Offset* is a time range and *Time* is an **enumeration** with values of *Dispatch, Start, Completion, Deadline,* and *NoIO* (i.e, of type *IO_Reference_Time*). *NoIO* indicates that the port does not produce output (e.g., in a given mode). If no value is specified, *Offset* is zero and *Time* is *Dispatch*. Declaration: Input_Time: **list of** IO_Time_Spec => ([Time => Dispatch; Offset => 0.0 ns .. 0.0 ns;]) **applies to** (port); Property Association Example: *Input_Time =>([Offset => 0.0 ns .. 10.0 ns; Time => Start;], [Offset => 5.0ns .. 20.0 ns; Time => Dispatch;]);*
Output_Rate	A property that specifies the number of outputs per dispatch or per second of data, events, event data, or initiations of subprogram calls. It can be specified only for ports and is a property of type *Rate_Spec*. *Rate_Spec* is a record type consisting of three fields: *Value_Range, Rate_Unit,* and *Rate_Distribution*. *Value_Range* is a range of real number; *Rate_Unit* is either *PerSecond* or *PerDispatch*; and *Rate_Distribution* can be assigned the **enumeration** values of *Fixed* or values specified by a user (i.e., *project-specified*). If a rate and unit value are not specified, the values are (1.0..1.0) input per thread dispatch. If no rate distribution is specified, the value is *Fixed*. Declaration: Output_Rate: Rate_Spec => [Value_Range => 1.0 .. 1.0; Rate_Unit => PerDispatch; Rate_Distribution => Fixed;] **applies to** (port);

Table A-7: *Communication-Related Properties (continued)*

Property	Description
Output_Rate (continued)	Property Association Example: *Output_Rate => [Value_Range => 20.0 .. 80.0; Rate_Unit => PerSecond; Rate_Distribution => Fixed;];*
Output_Time	A property that specifies the time in terms of execution time when output becomes available. The output time can be expressed relative to execution time at completion or start of execution, or relative to the deadline. It can be specified only for a port and is a list of *IO_Time_Spec* values. *IO_Time_Spec* is a record type with two fields *Offset* and *Time*. *Offset* is a time range and *Time* is an **enumeration** with values of *Dispatch, Start, Completion, Deadline*, and *NoIO* (i.e, of type *IO_Reference_Time*). *NoIO* indicates that the port does not produce output (e.g., in a given mode). If no value is specified, *Offset* is zero and *Time* is *Completion*. For data ports with a delayed connection, if no value is specified, the value for the field *Time* is *Deadline*. Declaration: Output_Time: **list of** IO_Time_Spec => ([Time => Completion; Offset => 0.0 ns .. 0.0 ns;]) **applies to** (port); Property Association Example: *Output_Time =>([Offset => 0.0 ns .. 10.0 ns; Time => Completion;], [Offset => 5.0 ns .. 20.0 ns; Time => Start;]);*
Subprogram_Call_Rate	A property that specifies the number of subprogram calls per dispatch or per second. It can be specified only for ports and is a property of type *Rate_Spec*. *Rate_Spec* is a record type consisting of three fields: *Value_Range, Rate_Unit*, and *Rate_Distribution*. *Value_Range* is a range of real number; *Rate_Unit* is either *PerSecond* or *PerDispatch*; and *Rate_Distribution* can be assigned the **enumeration** values of *Fixed* or values specified by a user (i.e., *project-specified*). If a rate and unit value are not specified, the values are (1.0..1.0) input per thread dispatch. If no rate distribution is specified, the value is *Fixed*. Declaration: Subprogram_Call_Rate: Rate_Spec => [Value_Range => 1.0 .. 1.0; Rate_Unit => PerDispatch; Rate_Distribution => Fixed;] **applies to** (subprogram access); Property Association Example: *Subprogram_Call_Rate => [Value_Range => 1.0 .. 2.0; Rate_Unit => PerSecond; Rate_Distribution => Fixed;];*

continues

Table A-7: *Communication-Related Properties (continued)*

Property	Description

Transmission_Time	A property that specifies the parameters for a linear model for the time interval between the start and end of a transmission of a sequence of N bytes onto a bus. It can be specified only for a bus. It is a property of type record with two fields: *Fixed* and *PerByte*. Both fields are a range of time values. The transition time for a message of N Bytes over a bus is determined by the values of this property when applied to the formula *Fixed* + N * *PerByte*. The range determines to lower and upper bound of the transmission time. The transmission time is the time between the transmission of the first bit of the message onto a bus and the transmission of the last bit of the message onto that bus. This time excludes arbitration, queuing, or any other times that are a function of how bus contention and scheduling are performed.
	
	Declaration:
	Transmission_Time: **record** (
	    Fixed: Time_Range;
	    PerByte: Time_Range;)
	    applies to (bus);
	
	Property Association Example:
	Transmission_Time => [Fixed => 1.0 ns .. 2.0 ns; PerByte => 5 ns .. 6 ns;];
	Corresponding Transmission Time value for a message of 10 Bytes:
	Transmission Time is the range: 51 ns .. 62 ns

Actual_Latency	A property that specifies the actual latency as determined by the implementation of an end-to-end flow through semantic connections. It can be specified for flows, connections, buses, devices, processors, and ports, and is a range of *Time* values (integer plus a time unit) that must be positive. The standard units are ns (nanoseconds), us (microseconds), ms (milliseconds), sec (seconds), min (minutes), and hr (hours). Declaration: Actual_Latency: Time_Range **applies to** (flow, connection, bus, device, processor, port); Property Association Example: *Actual_Latency => 5 ms .. 10 ms;*

Table A-7: *Communication-Related Properties (continued)*

Property	Description
Latency	A property that specifies the minimum and maximum amount of elapsed time allowed between the time the data or events enter the connection or flow and the time it exits. It can be specified for flows, connections, buses, devices, processors, and ports. It is a range of *Time* values (integer plus a time unit) that must be positive. The standard units are ns (nanoseconds), us (microseconds), ms (milliseconds), sec (seconds), min (minutes), and hr (hours). Declaration: Latency: Time_Range **applies to** (flow, connection, bus, device, processor, port); Property Association Example: *Latency => 2 ns .. 12 ns;*

A.5.5 Memory-Related Properties

Table A-8 lists properties useful in specifying data storage and memory and device access. The table includes property declarations taken from the standard and examples of property associations. Since all standard and user-defined properties that can be specified for at least one component of any category can be specified for an abstract component, the abstract component category is not included in the table.

Table A-8: *Memory-Related Properties*

Property	Description
Access_Right	A property that specifies the form of access permitted for a component. It can be specified for data, buses, data accesses, and bus accesses and is of property type *Access_Rights*. *Access_Rights* is an enumeration type with the values of *read_only, write_only, read_write,* and *by_method*. If no value is assigned to the property, the value is *read_write*. If the property is associated with a requires access it specifies the intended access to the component being accessed. If associated with a provides access it specifies the type of access that is permitted to the component for which access is provided. This access may be direct through read and write access or indirect through subprograms provided with the data type. The provided access specified

continues

Table A-8: *Memory-Related Properties (continued)*

Property	Description
Access_Right *(continued)*	by the value of *Access_Right* must not exceed the access right specified for the component itself. The required access specified by the value of *Access_Right* must not exceed the access right specified by the provides access or by the component itself. Declaration: Access_Right : Access_Rights => read_write **applies to** (data, bus, data access, bus access); Property Association Example: Access_Right => read_write;
Access_Time	A property that specifies the ranges of execution times during which a data component is being accessed. It can be specified only for data accesses. It is of type **record** with two fields: *First* and *Last*. Each of the fields is of type *IO_Time_Spec*. *IO_Time_Spec* is a **record** type with two fields: *Offset* and *Time*. *Offset* is of type *IO_Reference_Time* that is an **enumeration** type with values of *Dispatch, Start, Completion, Deadline,* and *NoIO*. *Time* is of type *Time_Range* that is a range of **aadlinteger** values with time units such as ms, ns, sec, and so on. If the property is not specified for a data access the value of each field is First => (Time => Start; Offset => 0.0 ns .. 0.0 ns;) and Last => (Time => Completion; Offset => 0.0 ns .. 0.0 ns;). Declaration: Access_Time: **record** (First: IO_Time_Spec ; Last: IO_Time_Spec ;) => [First =>[Time => Start; Offset => 0.0 ns .. 0.0 ns;]; Last => [Time => Completion; Offset => 0.0 ns .. 0.0 ns;];] **applies to** (data access); Property Association Example: *Access_Time => [First=>[Time=>Start; Offset => 0.0 ns .. 2.0 ns;]; Last=>[Time=>Completion; Offset=. 0.0 ns .. 2.0 ns;];] ;*

Table A-8: *Memory-Related Properties (continued)*

Property	Description
Allowed_Message_Size	A property that specifies the allowed range of sizes for a block of data that can be transmitted by a bus in a single transmission, when there is no packetization. It can be specified only for buses and is of type *Size_Range*. *Size_Range* is a range values of **aadlinteger** values with units of bits, Bytes, MByte => KByte, GByte, and TByte (i.e., a range of values of type *Size*). The values define the range of data message sizes—excluding any header or packetization overheads added due to bus protocols—that can be sent in a single transmission. Messages whose sizes fall below this range must be padded to fit the specified range. Messages whose sizes fall above this range must be broken into two or more separately transmitted packets. Declaration: Allowed_Message_Size: Size_Range **applies to** (bus); Property Association Example: *Allowed_Message_Size => 4 Bits .. 5 Bytes;*

Assign_Time	A property that specifies the values used in linear estimation of the execution time required to move a block of bytes on a particular processor. It can be specified only for processors. It is a property of type of **record** with two fields: *Fixed* and *PerByte*. Both of these fields are time ranges. The time range for the move is determined by multiplying the number of Bytes in a block times the value of *PerByte* plus the value of *Fixed*: (Number_of_Bytes * *PerByte*) + *Fixed*. The addition is an addition of both bounds of each range, yielding a new range value.
	
	Declaration:
	Assign_Time: **record** (
	   Fixed: Time_Range;
	   PerByte: Time_Range;)
	  applies to (processor);
	
	Property Association Example:
	Assign_Time => [Fixed=> 2 ns..3ns; PerByte=5ns..8ns;];

continues

Table A-8: *Memory-Related Properties (continued)*

Property	Description
Base_Address	A property that specifies the address of the first word in memory. It can be specified for memory, data, data accesses, and ports. It is a property of type **aadlinteger** with a maximum allowed value of *Max_Base_Address*. When specified for a memory component, it indicates the starting address for that memory. In the case of data components, data accesses, or ports it indicates the starting address of the memory to which they are bound. Declaration: Base_Address: **aadlinteger** 0 .. Max_Base_Address **applies to** (memory, data, data access, port); Property Association Example: *Base_Address => 1001;*
Device_Register_ Address	A property that specifics the address of a device register that is represented by a port declared for that device. It can be specified for ports and feature groups. It is a property of type **aadlinteger**. This property is optional, since ports may be represented by a source text variable within the device driver software. Declaration: Device_Register_Address: **aadlinteger applies to** (port, feature group); Property Association Example: *Device_Register_Address => 501;*

Read_Time	A property that specifies the values used in a linear estimation of the execution time required to read a block of bytes from memory. It can be specified only for memory. It is a property of type **record** with two fields: *Fixed* and *PerByte*. Both of these fields are time ranges. The time range for the read is determined by multiplying the number of Bytes in a block times the value of *PerByte* plus the value of *Fixed*: (Number_of_Bytes * *PerByte*) + *Fixed*. The addition is an addition of both bounds of each range, yielding a new range value.
	
	Declaration:
	Read_Time: **record** (
	   Fixed: Time_Range;
	   PerByte: Time_Range;)
	   applies to (memory);

Table A-8: *Memory-Related Properties (continued)*

Property	Description
Read_Time *(continued)*	Property Association Example: *Read_Time => [Fixed=> 2 ns..3ns; PerByte=5ns..8ns;];*
Source_Code_Size	A property that specifies the size of the static code and read-only data that results when associated source text is compiled, linked, bound, and loaded in a final system. It can be specified for data, threads, thread groups, processes, systems, subprograms, processors, and devices. The property type is *Size* that is an integer with units of *bits*, *Bytes* (bytes), *KByte* (kilobytes), *MByte* (megabytes), and *GByte* (gigabytes). Declaration: Source_Code_Size: Size **applies to** (data, thread, thread group, process, system, subprogram, processor, device); Property Association Example: *Source_Code_Size => 4 MByte;*
Source_Data_Size	A property that specifies the size of the readable and writeable data that results when associated source text is compiled, linked, bound, and loaded in the final system. In the case of data types, it specifies the maximum size required to hold a value of an instance of the data type. It can be specified for data, subprograms, threads, thread groups, processes, systems, processors, and devices. The property type is *Size* that is an integer with units of bits, Bytes (bytes), KByte (kilobytes), MByte (megabytes), and GByte (gigabytes). Declaration: Source_Data_Size: Size **applies to** (data, subprogram, thread, thread group, process, system, processor, device); Property Association Example: *Source_Data_Size => 1 MByte;*
Source_Heap_Size	A property that specifies the minimum and maximum heap size requirements of a thread or subprogram. It can be specified for threads and subprograms. The property type is *Size* that is an integer with units of bits, Bytes (bytes), KByte (kilobytes), MByte (megabytes), and GByte (gigabytes).

continues

Table A-8: *Memory-Related Properties (continued)*

Property	Description
Source_Heap_Size *(continued)*	Declaration: Source_Heap_Size: Size **applies to** (thread, subprogram); Property Association Example: *Source_Heap_Size => 1 MByte;*
Source_Stack_Size	A property that specifies the maximum size of the stack used by a processor executive, a device driver, a thread, or a subprogram during execution. It can be specified for threads, subprograms, processors, and devices. The property type is *Size* that is an integer with units of bits, Bytes (bytes), KByte (kilobytes), MByte (megabytes), and GByte (gigabytes). Declaration: Source_Stack_Size: Size **applies to** (thread, subprogram, processor, device); Property Association Example: *Source_Stack_Size => 25 KByte;*
Byte_Count	A property that specifies the number of bytes in memory. It can be specified only for memory and is an integer value. The maximum value that can be specified is the value of the property *Max_Byte_Count*. Declaration: Byte_Count: **aadlinteger** 0 .. Max_Byte_Count **applies to** (memory); Property Association Example: *Byte_Count => 1001;*
Word_Size	A property that specifies the size of the smallest independently readable and writeable unit of storage in the memory. It can be specified only for memory. The property type is *Size* that is an integer with units of bits, Bytes (bytes), KByte (kilobytes), MByte (megabytes), and GByte (gigabytes). If no value is specified, the value is *8 bits*.

Table A-8: *Memory-Related Properties (continued)*

Property	Description
Word_Size (continued)	Declaration: Word_Size: Size => 8 bits **applies to** (memory); Property Association Example: *Word_Size => 64 Bytes;*
Word_Space	A property that specifies the interval between successive addresses used for successive words of memory. It can be specified only for memory and is an integer value. The maximum value that can be specified is the value of the property *Max_Word_Space*. A value greater than 1 means the addresses used to access words are not contiguous integers. If no value is specified, the value is 1. Declaration: Word_Space: **aadlinteger** 1 .. Max_Word_Space => 1 **applies to** (memory); Property Association Example: *Word_Size => 64 Bytes;*

Write_Time	A property that specifies the values used in a linear estimation of the execution time required to write a block of bytes to memory. It can be specified only for memory. It is a property of type **record** with two fields: *Fixed* and *PerByte*. Both of these fields are time ranges. The time range for the read is determined by multiplying the number of Bytes in a block times the value of *PerByte* plus the value of *Fixed*: (Number_of_Bytes * *PerByte*) + *Fixed*. The addition is an addition of both bounds of each range, yielding a new range value.
	
	 Declaration:
	 Write_Time: **record** (
	   Fixed: Time_Range;
	   PerByte: Time_Range;)
	   applies to (memory);
	
	 Property Association Example:
	 Write_Time => (Fixed=> 2 ns..3ns; PerByte=5ns..8ns;);

A.5.6 Programming Properties

Table A-9 lists properties useful in specifying information related to the representation and implementation of application and hardware components using an implementation language. The table includes property declarations taken from the standard and examples of property associations. Since all standard and user-defined properties that can be specified for at least one component of any category can be specified for an abstract component, the abstract component category is not included in the table.

Table A-9: *Programming Properties*

Property	Description
Activate_Entrypoint	A property that specifies a subprogram classifier. It can be specified for threads and devices and is a property of type **classifier**. The specified classifier represents a subprogram in source text that will execute when a thread or device is activated as part of a mode switch. The subprogram must be visible in and callable from the outermost program scope, as defined by the scope and visibility rules of the applicable implementation language. The source language annex of the AADL standard defines acceptable parameter and result signatures for the entry point subprogram. Declaration: Activate_Entrypoint: **classifier** (subprogram classifier) **applies** to (thread, device); Property Association Example: *Activate_Entrypoint => rt_init.impl;*
Activate_Entrypoint_ Call_Sequence	A property that specifies a subprogram call sequence that executes after a thread has been dispatched. It can be specified for threads and devices and is a property of type **reference**. If the property is specified for a provided subprogram, event port, or event data port feature, this entrypoint is chosen when the corresponding call, event, or event data arrives instead of the compute entrypoint specified for the containing thread. The call sequence must exist in the source text as a parameter-less function that performs the calls and passes the appropriate port and parameter values as actual parameters. This function may be generated from AADL. This function must be visible in and callable from the outermost program scope, as specified by the relevant rules of the applicable source language. The source language annex

Table A-9: *Programming Properties (continued)*

Property	Description
Activate_Entrypoint_ Call_Sequence (continued)	of the AADL standard defines acceptable parameters and result signatures for the entrypoint subprogram. Declaration: Activate_Entrypoint_Call_Sequence: **reference** (subprogram call sequence) **applies to** (thread, device); Property Association Example: *Activate_Entrypoint_Call_Sequence => reference (basic_sequence);*
Activate_Entrypoint_ Source_Text	A property that specifies the name of the source text code that can execute when a thread is activated. It can be specified for threads and devices and is a property of type **aadlstring**. The source text must be visible in and callable from the outermost program scope, as specified by the relevant rules of the applicable source language. The source language annex of the AADL standard defines acceptable parameters and result signatures for the entrypoint subprogram. Declaration: Activate_Entrypoint_Source_Text: **aadlstring applies to** (thread, device); Property Association Example: *Activate_Entrypoint_Source_Text => "act_ep_code.ccp";*
Compute_Entrypoint	A property that specifies the subprogram classifier of a subprogram in source text that is executed after a thread has been dispatched. It can be specified for threads and devices and provides subprogram accesses, event ports, and event data ports. It is a property of type **classifier**. If specified for a provided subprogram, event port, or event data port feature, this entrypoint is chosen when the corresponding call, event, or event data arrives instead of the specified compute entrypoint for the containing thread. The subprogram in the source text must be visible in and callable from the outermost program scope, as specified by the relevant rules of the applicable source language. The source language annex of the AADL standard defines acceptable parameters and result signatures for the entrypoint subprogram.

continues

Table A-9: *Programming Properties (continued)*

Property	Description
Compute_Entrypoint *(continued)*	Declaration: Compute_Entrypoint: **classifier** (subprogram classifier) **applies to** (thread, device, provides subprogram access, event port, event data port); Property Association Example: Compute_Entrypoint => *classifier (get_param.basic);*
Compute_Entrypoint_ Call_Sequence	A property that specifies the subprogram call sequence that can execute after a thread has been dispatched. It can be specified for threads and devices and provides subprogram accesses, event ports, and event data ports. If specified for a provided subprogram, event port, or event data port feature, the entrypoint is chosen when the corresponding call, event, or event data arrives instead of the compute entrypoint specified for the containing thread. The call sequence must exist in the source text as a parameterless function that performs the calls and passes the appropriate port and parameter values as actual parameters. This function must be visible in and callable from the outermost program scope, as specified by the relevant rules of the applicable source language. The source language annex of the AADL standard defines acceptable parameters and result signatures for the entrypoint subprogram. Declaration: Compute_Entrypoint_Call_Sequence: **reference** (subprogram call sequence) **applies to** (thread, device, provides subprogram access, event port, event data port); Property Association Example: *Compute_Entrypoint_Call_Sequence => **reference** (compute_ep_seq);*
Compute_Entrypoint_ Source_Text	A property that specifies the name of the source text code that can execute when a thread is activated. It can be specified for threads, devices, provides subprogram accesses, event ports, and event data ports and is a property of type **aadlstring**. If the property is specified for a provided subprogram, event port, or event data port feature, this entrypoint is chosen when the corresponding call, event, or event data arrives instead of the compute entrypoint specified for the containing thread. The code sequence in the source text must be visible in and callable from the outermost program scope, as specified by the relevant rules of the applicable source language. The source language annex of the AADL standard defines acceptable parameters and result signatures for the entrypoint subprogram.

Table A-9: *Programming Properties (continued)*

Property	Description
Compute_Entrypoint_Source_Text (continued)	Declaration: Compute_Entrypoint_Source_Text: **aadlstring applies to** (thread, device, provides subprogram access, event port, event data port); Property Association Example: *Compute_Entrypoint_Source_Text => "compute_ep_code";*
Deactivate_Entrypoint	A property that specifies the subprogram classifier that represents the subprogram in source text that executes when a thread is deactivated as part of a mode switch. It can be specified for threads and devices. It is a property of type **classifier**. The subprogram in the source text must be visible in and callable from the outermost program scope, as specified by the relevant rules of the applicable source language. The source language annex of the AADL standard defines acceptable parameters and result signatures for the entry-point subprogram. Declaration: Deactivate_Entrypoint: **classifier** (subprogram classifier) **applies to** (thread, device); Property Association Example: *Deactivate_Entrypoint => **classifier** (shut_down);*
Deactivate_Entrypoint_Call_Sequence	A property that specifies the subprogram call sequence that executes when a thread is deactivated as part of a mode switch. It can be specified for threads and devices and is a property of type **reference**. The call sequence must exist in the source text as a parameterless function that performs the calls and passes the appropriate port and parameter values as actual parameters. This function must be visible in and callable from the outermost program scope, as specified by the relevant rules of the applicable source language. The source language annex of the AADL standard defines acceptable parameters and result signatures for the entry-point subprogram. Declaration: Deactivate_Entrypoint_Call_Sequence: **reference** (subprogram call sequence) **applies to** (thread, device); Property Association Example: *Deactivate_Entrypoint_Call_Sequence => **reference** (deactivate_ep_seq)*

continues

Table A-9: *Programming Properties (continued)*

Property	Description
Deactivate_Entrypoint_ Source_Text	A property that specifies the name of a source text code sequence that executes when a thread is deactivated. It can be specified only for threads and is a property of type **aadlstring**. The code sequence must be visible in and callable from the outermost program scope, as specified by the relevant rules of the applicable source language. The source language annex of the AADL standard defines acceptable parameters and result signatures for the entrypoint subprogram. Declaration: Deactivate_Entrypoint_Source_Text: **aadlstring applies to** (thread); Property Association Example: *Deactivate_Entrypoint_Source_Text => "deactivate_ep_code";*
Finalize_Entrypoint	A property that specifies the subprogram classifier that represents the subprogram in source text that executes when a thread is finalized. It can be specified for threads and devices and is a property of type **classifier**. The subprogram in the source text must be visible in and callable from the outermost program scope, as specified by the relevant rules of the applicable source language. The source language annex of the AADL standard defines acceptable parameters and result signatures for the entrypoint subprogram. Declaration: Finalize_Entrypoint: **classifier** (subprogram classifier) **applies to** (thread, device); Property Association Example: Finalize _Entrypoint => **classifier** *(finalize_ep);*
Finalize_Entrypoint_ Call_Sequence	A property that specifies the name of a subprogram call sequence that executes when a thread is finalized. It can be specified only for threads and devices and is a property of type **reference**. The call sequence must exist in the source text as a function without parameters that performs the calls and passes the appropriate port and parameter values as actual parameters. This function must be visible in and callable from the outermost program scope, as specified by the relevant rules of the applicable source language. The source language code generation annex of the AADL standard defines acceptable parameters and signatures for the entrypoint subprogram.

Table A-9: *Programming Properties (continued)*

Property	Description
Finalize_Entrypoint_ Call_Sequence (continued)	Declaration: Finalize_Entrypoint_Call_Sequence: **reference** (subprogram call sequence) **applies to** (thread, device); Property Association Example: *Finalize_Entrypoint_Call_Sequence => **reference** (finalize_ep_seq);*
Finalize_Entrypoint_ Source_Text	A property that specifies the source text code sequence that executes when a thread is finalized. It can be specified only for threads and devices, and is a property of type **aadlstring**. The code sequence must be visible in and callable from the outermost program scope, as specified by the relevant rules of the applicable source language. The source language annex of the AADL standard defines acceptable parameters and signatures for the entrypoint subprogram. Declaration: Finalize_Entrypoint_Source_Text: **aadlstring applies to** (thread, device); Property Association Example: *Finalize_Entrypoint_Source_Text => "finalize_ep_code";*
Initialize_Entrypoint	A property that specifies the subprogram classifier that represents the subprogram in source text that executes when a thread is initialized. It can be specified for threads and devices and is a property of type **classifier**. The subprogram in the source text must be visible in and callable from the outermost program scope, as specified by the relevant rules of the applicable source language. The source language annex of the AADL standard defines acceptable parameters and result signatures for the entrypoint subprogram. Declaration: Initialize_Entrypoint: **classifier** (subprogram classifier) **applies to** (thread, device); Property Association Example: *Initialize_Entrypoint => **classifier** (controller_init);*
Initialize_Entrypoint_ Call_Sequence	A property that specifies the subprogram call sequence that executes when a thread is initialized. It can be specified for threads and devices and is a property of type **reference**. The call sequence must exist in the source text as a parameter-less function that performs the calls and passes the appropriate port and parameter

continues

Table A-9: *Programming Properties (continued)*

Property	Description
Initialize_Entrypoint_Call_Sequence (continued)	values as actual parameters. This function must be visible in and callable from the outermost program scope, as specified by the relevant rules of the applicable source language. The source language annex of the AADL standard defines acceptable parameters and result signatures for the entrypoint subprogram. Declaration: Initialize_Entrypoint_Call_Sequence: **reference** (subprogram call sequence) **applies to** (thread, device); Property Association Example: *Initialize_Entrypoint_Call_Sequence => **reference** (initialize_ep_seq);*
Initialize_Entrypoint_Source_Text	A property that specifies the name of a source text code sequence that executes when a thread is to be initialized. It can be specified only for threads and devices, and is a property of type **aadlstring**. The code sequence must be visible in and callable from the outermost program scope, as specified by the relevant rules of the applicable source language. The source language annex of the AADL standard defines acceptable parameters and signatures for the entrypoint subprogram. Declaration: Initialize_Entrypoint_Source_Text: **aadlstring applies to** (thread, device); Property Association Example: *Initialize_Entrypoint_Source_Text => "initialize_ep_code";*
Recover_Entrypoint	A property that specifies the name of a subprogram classifier representing a subprogram in source text that executes when a thread is in recovery. It can be specified for threads and devices and is a property of type **classifier**. The subprogram in the source text must be visible in and callable from the outermost program scope, as specified by the relevant rules of the applicable source language. The source language annex of the AADL standard defines acceptable parameters and signatures for the entrypoint subprogram. Declaration: Recover_Entrypoint: **classifier** (subprogram classifier) **applies to** (thread, device); Property Association Example: *Recover_Entrypoint => **classifier** (controller_recov.impl);*

Table A-9: *Programming Properties (continued)*

Property	Description
Recover_Entrypoint_ Call_Sequence	A property that specifies the subprogram call sequence that executes when a thread is in recovery. It can be specified for threads and devices and is a property of type **reference**. The call sequence must exist in the source text as a parameter-less function that performs the calls and passes the appropriate port and parameter values as actual parameters. This function must be visible in and callable from the outermost program scope, as specified by the relevant rules of the applicable source language. The source language annex of the AADL standard defines acceptable parameters and result signatures for the entrypoint subprogram. Declaration: Recover_Entrypoint_Call_Sequence: **reference** (subprogram call sequence) **applies to** (thread, device); Property Association Example: *Recover_Entrypoint_Call_Sequence => **reference** (recover_ep_seq);*
Recovery_Entrypoint_ Source_Text	A property that specifies the name of a source text code sequence that executes when a thread is recovering from a fault. It can be specified for threads and devices, and is a property of type **aadlstring**. The code sequence must be visible in and callable from the outermost program scope, as specified by the relevant rules of the applicable source language. The source language annex of the AADL standard defines acceptable parameters and signatures for the entrypoint subprogram. Declaration: Recover_Entrypoint_Source_Text: **aadlstring applies to** (thread, device); Property Association Example: *Recover_Entrypoint_Source_Text => "initialize_ep_code";*
Source_Language	An inherited property that specifies the applicable programming language. It can be specified for subprograms, data, threads, thread groups, processes, systems, buses, devices, virtual buses, virtual processors, and processors. It is a list of supported programming languages. The supported languages can be defined by a user. These are defined as values of the **enumeration** property *Supported_ Source_Languages*. Example values that a user may choose to define include *Ada95, Ada2005, C,* and *Simulink_6_5.*

continues

Table A-9: *Programming Properties (continued)*

Property	Description
Source_Language *(continued)*	When specified for a component, the source text associated with that component and the source language and source text specified for all of its software subcomponents must comply with the applicable language standard. Where a source language is not specified, a tool may infer a source language from the file extension. A tool may establish a default value for the source language property. Declaration: Source_Language: **inherit list of** Supported_Source_Languages **applies to** (subprogram, data, thread, thread group, process, system, bus, device, processor, virtual bus, virtual processor); Property Association Examples: *Source_Language => (C);* *Source_Language => (Ada95, Ada2005, C);*
Source_Name	A property that specifies a source declaration or source name within associated source text that corresponds to a feature identifier. It can be specified for data, ports, subprograms, virtual buses, virtual processors and parameters. Declaration: Source_Name: **aadlstring applies to** (data, port, subprogram, parameter, virtual bus, virtual processor); Property Association Example: *Source_Name => "GPS_data";*
Source_Text	An inherited property that specifies a list of files that contain source text. It can be specified for data, ports, subprograms, threads, thread groups, processes, systems, memory, buses, devices, processors, parameters, feature groups, virtual buses, virtual processors, and packages. It is a list of string values. The combined source text contained in all files listed must form one or more separately compileable units as defined in the applicable source language standard. Declaration: Source_Text: **inherit list of aadlstring applies to** (data, port, subprogram, thread, thread group, process, system, memory, bus, device, processor, parameter, feature group, virtual bus, virtual processor, package);

Table A-9: *Programming Properties (continued)*

Property	Description
Source_Text (continued)	Property Association Examples: *Source_Text => ("control_src.cpp");* *Source_Text => ("control_src.cpp", "interface.cpp", "montor.cpp");* The Device_Driver property specifies a reference to an abstract component implementation that contains an instance of a driver subprogram or an instance of an (ISR) thread. Declaration: Device_Driver: **classifier** (abstract implementation) **applies to** (device);
Supported_Source_ Language	A property that specifies the source language(s) supported by a processor. It can be specified for processors, virtual processors, and systems and is a list of the supported programming languages. The supported languages are *project-specified* and can be defined by a user. These are defined as values of the **enumeration** property *Supported_Source_Languages*. Example values that a user may choose to define include *Ada95, Ada2005, C,* and *Simulink_6_5.* If an allowed source language list is not specified, there are no restrictions on the source. The source language of every software component that may be accessed by any thread bound to a processor must appear in the list of allowed source languages for that processor. Declaration: Supported_Source_Language: **list of** Supported_Source_Languages **applies to** (processor, virtual processor, system); Property Association Examples: *Supported_Source_Language => (C);* *Supported_Source_Language => (Ada95, Ada2005, C)*
Hardware_Description_ Source_Text	An inherited property that specifies a list of files that contain hardware description language source text. It can be specified for memory, buses, devices, processors, and systems and is a list of string values. Each string value is interpreted as a POSIX pathname and must satisfy the syntax and semantics for path names as defined in the POSIX standard. Extensions to the standard POSIX pathname syntax and semantics are permitted. For example, environment variables

continues

Table A-9: *Programming Properties (continued)*

Property	Description
Hardware_Description_Source_Text (continued)	and regular expressions are permitted. Special characters may be used to assist in configuration management and version control. If the first character of a pathname is . (dot), the path is relative to the directory in which the file containing the AADL specification text is stored. Otherwise, the tool may define the default directory used to locate the file designated by a pathname. Declaration: Hardware_Description_Source_Text: **inherit list of aadlstring applies to** (memory, bus, device, processor, system); Property Association Examples: *Hardware_Description_Source_Text => ("Hi_speed_Processor.vhdl");* *Hardware_Description_Source_Text => (" Hspd.vhdl ", "Core_Proc.vhdl");*
Hardware_Source_Language	A property that specifies a list of supported hardware description languages. It can be specified for memory, buses, devices, processors, and systems. The supported languages are *project-specified* and can be defined by a user. These are defined as values of the **enumeration** property *Supported_Hardware_Source_Languages*. Declaration: Hardware_Source_Language: Supported_Hardware_Source_Languages **applies to** (memory, bus, device, processor, system); Property Association Example: *Hardware_Source_Language => VHDL;*

A.5.7 Modeling Properties

Properties useful in specifying an AADL model are listed in Table A-10. The table includes property declarations taken from the standard and examples of property associations. Since all standard and user-defined properties that can be specified for at least one component of any category can be specified for an abstract component, the abstract component category is not included in the table.

Table A-10: *Some Useful Modeling Properties (continued)*

Table A-10: *Some Useful Modeling Properties*

Property	Description
Acceptable_Array_Size	A property that specifies a set of closed memory size ranges for an array. It can be specified for subcomponents and features. Each range includes a lower and an upper bound with a maximum value of *Max_Memory_Size*. It is a property of type **list of** *Size_Range* where each entry in the list represents the range for each dimension of an array. The minimum and maximum values for each range are of type *Size* (i.e., an integer with memory size units). The standard memory size units are bits, Bytes, KByte, MByte, GByte, and TByte. Declaration: Acceptable_Array_Size: **list of** Size_Range **applies to** (subcomponent, feature); Property Association Examples: *Acceptable_Array_Size => (0 Bytes..10 Bytes, 2Bytes..5Bytes, 5Bytes..10Bytes);* *Acceptable_Array_Size => (0 MByte..10 MByte)*
Classifier_Matching_ Rule	An inherited property that specifies acceptable matches of classifiers between the source and the destination of a connection. It can be specified for component implementations and connections. The property is an **enumeration** type with values of *Classifier_Match, Equivalence, Subset, Conversion, and Complement.* Declaration: Classifier_Matching_Rule: **inherit enumeration** (Classifier_Match, Equivalence, Subset, Conversion, Complement) **applies to** (connection, component implementation); Property Association Example: *Classifier_Matching_Rule => Classifier_Match;*
Classifier_Substitution_ Rule	An inherited property that specifies acceptable substitutions of classifiers in a **refined to** declaration. It can be specified for classifiers and subcomponents. The property is an **enumeration** type with values of *Classifier_Match, Type_Extension,* and *Signature_Match.* Declaration: Classifier_Substitution_Rule: **inherit enumeration** (Classifier_ Match, Type_Extension, Signature_Match) **applies to** (classifier, subcomponent);

continues

Table A-10: *Some Useful Modeling Properties (continued)*

Property	Description
Classifier_Substitution_ Rule (continued)	Property Association Example: *Classifier_Substitution_Rule: => Classifier_Match;*
Implemented_As	A property that specifies a system implementation that describes how the internals of an execution platform component are realized. It can be specified for memory, buses, virtual buses, devices, virtual processors, and systems. This allows systems to be modeled as a layered architecture using the execution platform as a layering abstraction. The property values are system implementation classifiers. Declaration: Implemented_As: **classifier** (system implementation) **applies to** (memory, bus, virtual bus, device, virtual processor, processor, system); Property Association Example: *Implemented_As: =>* **classifier** *(flight_controller.basic);*
Prototype_Substitution_ Rule	An inherited property that specifies acceptable classifiers that can be supplied as actual in a prototype binding. It can be specified for prototypes and classifiers. The property is an **enumeration** type with values of *Classifier_Match, Type_Extension,* and *Signature_Match.* Declaration: Prototype_Substitution_Rule: **inherit enumeration** (Classifier_Match, Type_Extension, Signature_Match) **applies to** (prototype, classifier); Property Association Example: *Prototype_Substitution_Rule: => Classifier_Match;*

A.5.8 Project-Specific Constants and Property Types

The AADL_Project property set is a modifiable collection of enumeration property types and property constants that you can customize to create a customized project-specific set. It is part of every AADL specification and must be provided by AADL tool suppliers.

A description of the content of the AADL_Project property set is provided in Table A-11. The entries *project-specified* and *project-specified-integer-literal* in a type declaration indicate that actual values are to be supplied by the person using the property set. The values do not include the angular bracket (< >) symbols.

Table A-11: *The AADL_Project Property Set*

Property	Description
Supported_Active_ Thread_Handling_ Protocols	An enumeration property type that specifies the set of possible actions that can be taken to handle threads that are in the state of performing computation at the instant of a mode switch. A single value *abort* is declared in the standard. Others are designated as *project-specified* and can be defined by a user. Some values that may be added include *suspend, complete_one, or complete_all.* Declaration: Supported_Active_Thread_Handling_Protocols: **type enumeration** (abort, *<project-specified>*);
Supported_Connection_ Patterns	An enumeration property type that specifies the set of patterns that are supported in connecting a source component array and a destination component array. The values defined for this numeration type are *One_To_One* represents the case when each item of the source component array is connected to the corresponding item of the destination component array. This property can be used only if the source and the destination component arrays have the same dimension and same dimension size. *All_To_All* represents the situation where each item of the source component array is connected to each item of the destination component array. *Next* or *Previous* indicates that elements of the ultimate source array dimension are connected to the next (previous) element in the ultimate destination array dimension without wrapping between the first and last. *Cyclic_Next* or *Cyclic_Previous* value indicates that elements of the ultimate source array dimension are connected to the next (previous) element in the ultimate destination array dimension. *One_to_All* indicates that a single element of the ultimate source has a semantic connection to each element in the ultimate destination. *All_to_One* indicates that each array element of the ultimate source has a semantic connection to a single element in the ultimate destination. Declaration: Supported_Connection_Patterns: **type enumeration** (One_To_One, All_To_All, One_To_All, All_To_One, Next, Previous, Cyclic_Next, Cyclic_Previous);

continues

Table A-11: *The AADL_Project Property Set (continued)*

Property	Description
Supported_ Concurrency_Control_ Protocols	An **enumeration** property type that specifies the set of concurrency control protocols that are supported. A single value (*None_Specified*) is declared in the Standard. Others are designated as *project-specified* and can be defined by a user. Some example values that a user may choose to add include *Interrupt_Masking, Maximum_Priority, Priority_Inheritance, Priority_Ceiling, Spin_Lock,* and *Semaphore*. Declaration: Supported_Concurrency_Control_Protocols: **type enumeration** (None_Specified, Interrupt_Masking, Maximum_Priority, Priority_Inheritance, Priority_Ceiling, Spin_Lock, Semaphore);
Supported_Dispatch_ Protocols	An **enumeration** property type that specifies the set of thread dispatch protocols that are supported. A single value *Periodic* is declared in the Standard. Others are designated as *project-specific* and can be defined by a user. Some example values that a user may choose to add include *Periodic, Sporadic, Aperiodic, Timed, Hybrid,* and *Background*. These protocols have semantics that are defined in the Standard and are described as follows: *Periodic* represents periodic dispatch of threads with deadlines. *Sporadic* represents event-triggered dispatching of threads with soft deadlines. *Aperiodic* represents event-triggered dispatch of threads with hard deadlines. *Timed* represents threads that are dispatched after a given time unless they are dispatched by arrival of an event or event data. *Hybrid* represents threads that are dispatched by both an event or event data arrival and periodically. *Background* represents threads that are dispatched once and execute until completion. A *Period* must be defined for *Periodic, Timed,* and *Hybrid* threads. Declaration: Supported_Dispatch_Protocols: **type enumeration** (Periodic, Sporadic, Aperiodic, Timed, Hybrid, Background);
Supported_Queue_ Processing_Protocols	An enumeration property type that specifies the set of queue processing protocols that are supported. A single value (*FIFO*) that represents the first-in, first-out processing of queues is declared in the Standard. Others are designated as *project-specified* and can be defined by a user.

Table A-11: *The AADL_Project Property Set (continued)*

Property	Description
Supported_Queue_ Processing_Protocols (continued)	Declaration: Supported_Queue_Processing_Protocols: **type enumeration** (FIFO, *<project-specified>*);
Supported_Hardware_ Source_Languages	An enumeration property type that specifies the set of hardware description languages that are supported. The values are *project-specified* and can be defined by a user. An example value that a user may choose to add is the *VHDL* hardware description language. Declaration: Supported_Hardware_Source_Languages: **type enumeration** (VHDL,*<project-specified>*);
Supported_Connection_ QoS	An enumeration property type that specifies the quality of services supported. The values *GuaranteedDelivery*, *OrderedDelivery*, and *SecureDelivery* are defined and *project-specified* values can be defined by a user. Declaration: Supported_Connection_QoS : **type enumeration** (GuaranteedDelivery, OrderedDelivery, SecureDelivery, <project specific>);

Supported_Scheduling_ Protocols	An enumeration property type that specifies the set of scheduling protocols that are supported. *project-specified* values can be defined by a user. Some example values that a user may choose to add include *FixedTimeline, Cooperative, RMS, EDF, SporadicServer, SlackServer,* and *ARINC653*.
	The semantics for these protocols are summarized here.
	• None (single thread).
	• Interrupt-driven (handling of interrupt service routines (ISR)).
	• For periodic task sets: fixed timeline, cooperative (cyclic executive), deadline monotonic, least laxity.
	• For hybrid task set:
	  – Fixed priority server based on Rate Monotonic Scheduling (RMS) (polling server, deferrable server, sporadic server, slack stealer).
	  – Dynamic priority server based on Earliest Deadline First (EDF) (dynamic polling server, dynamic deferrable server, dynamic sporadic server, total bandwidth server, constant bandwidth server).

continues

Table A-11: *The AADL_Project Property Set (continued)*

Property	Description
Supported_Scheduling_ Protocols (continued)	Scheduling protocols have a policy for scheduling periodic, aperiodic/sporadic, and background threads. For example, in Rate-Monotonic-Scheduling (RMS), priority assignments for periodic threads are made according to decreasing rate, for aperiodic and sporadic threads according to their minimum interarrival time, and for background threads as FIFO. Declaration: Supported_Scheduling_Protocols: **type enumeration** (FixedTimeline, Cooperative, RMS, EDF, SporadicServer, SlackServer, ARINC653, *<project-specified>*);
Supported_Source_ Languages	An enumeration property type that specifies the set of software source languages that are supported. The values are *project-specified* and can be defined by a user. Example values that a user may choose to add include *Ada95, Ada2005, C,* and *Simulink_6_5.* Declaration: Supported_Source_Languages: **type enumeration** (Ada95, Ada2005, C, Simulink_6_5, *<project-specified>*);
Supported_Distributions	An enumeration property type that specifies the set of distribution functions that are supported. A single value *Fixed.* Others are designated as *project-specified* and can be defined by a user, for example *Poisson.* Declaration: Supported_Distributions: **type enumeration** (Fixed, Poisson, *<project-specified>*);
Supported_Classifier_ Substitutions	An enumeration property type that specifies the set of classifier substitutions that are supported for prototypes and refinement declarations. Three values *Classifier_Match, Type_Extension,* and *Signature_Match* are defined in the Standard and *project-specified* values can be defined by a user. For example, a signature matching with name mapping can be added. Declaration: Supported_Classifier_Substitutions: **type enumeration** (Classifier_Match, Type_Extension, Signature_Match, *<project-specified>*);

Table A-11: *The AADL_Project Property Set (continued)*

Property	Description

Data_Volume	A property type that defines data volume as data per time units. This is an integer set with units expressed as *Size* units per second of time (*ps*) resulting in the units *bitsps, Bytesps, KBytesps, MBytesps*, and *GBytesps*. The numeric value for a property of this type must be positive.
	
	Declaration:
	Data_Volume: **type aadlinteger** 0 bitsps .. Max_Aadlinteger
	        units (bitsps, Bytesps => bitsps * 8,
	            KBytesps => Bytesps * 1000,
	            MBytesps => KBytesps * 1000,
	            GBytesps => MBytesps * 1000);

Max_Aadlinteger	An integer property constant that specifies the largest machine integer value that can be represented. It may be used as the maximum value in property associations but does not specify the maximum integer representation on a target processor. A value is not defined in the Standard; a *project-specified* value must be declared. Declaration: Max_Aadlinteger: **constant aadlinteger** => *<project-specified-integer-literal>*;
Max_Target_Integer	An integer property constant that specifies the largest machine integer value on a target processor. A value is not defined in the Standard; a *project-specified* value must be declared. Declaration: Max_Target_Integer: **constant aadlinteger** => *<project-specified-integer-literal>*;
Max_Base_Address	An integer property constant that specifies the largest the maximum value that can be declared for the *Base_Address* property. A value is not defined in the Standard; a *project-specified* value must be declared. Declaration: Max_Base_Address: **constant aadlinteger** => *<project-specified-integer-literal>*;

continues

Table A-11: *The AADL_Project Property Set (continued)*

Property	Description
Max_Memory_Size	A property constant of type *Size* that specifies the maximum memory size that can be declared (i.e., the maximum value of the property *Size*). A value is not defined in the Standard; a *project-specified* value must be declared. Declaration: Max_Memory_Size: **constant** Size => *<project-specified-aadl-integer>*;
Max_Queue_Size	An **aadlinteger** property constant that specifies the maximum value that can be declared for the *Queue_Size* property. A value is not defined in the Standard; a *project-specified* value must be declared. Declaration: Max_Queue_Size: **constant aadlinteger** => *<project-specified-integer-literal>*;
Max_Thread_Limit	An **aadlinteger** property constant that specifies the maximum value that can be declared for the *Thread_Limit* property. A value is not defined in the Standard; a *project-specified* value must be declared. Declaration: Max_Thread_Limit: **constant aadlinteger** => *<project-specified-integer-literal>*;
Max_Time	A property constant of type *Time* that specifies the maximum value that can be declared for a property of the type *Time*. A value is not defined in the Standard; a *project-specified* value must be declared. Declaration: Max_Time: **constant** Time => *<project-specified-integer-literal>*;
Max_Urgency	An **aadlinterger** property constant that specifies the maximum value that can be declared for the *Urgency* property. A value is not defined in the Standard; a *project-specified* value must be declared. Declaration: Max_Urgency: **constant aadlinteger** => *<project-specified-integer-literal>*;

Table A-11: *The AADL_Project Property Set (continued)*

Property	Description
Max_Byte_Count	An **aadlinterger** property constant that specifies the maximum value that can be declared for the *Byte_Count* property. A value is not defined in the Standard; a *project-specified* value must be declared. Declaration: Max_Byte_Count: **constant aadlinteger** => *<project-specified-integer-literal>*;
Max_Word_Space:	An **aadlinterger** property constant that specifies the maximum value that can be declared in for the *Word_Space* property. A value is not defined in the Standard; a *project-specified* value must be declared. Declaration: Max_Word_Space: **constant aadlinteger** => *<project-specified-integer-literal>*;
Size_Units	A set of units that defines a measurement of size that is available for use in other property definitions. Users may add to this units type. The conversion factor of 1000 is consistent with ISO/SI. *B, KB*, and so on in AADL 2004 have been replaced by Byte, Kbyte and so on in AADL V2. Since AADL is case insensitive, Kb could have been interpreted as K bits rather than K bytes. Declaration: Size_Units: **type units** (bits, Bytes => bits * 8, KByte => Bytes* 1000, MByte => KByte* 1000, GByte => MByte * 1000, TByte => GByte * 1000);
Time_Units	A set of units that defines a measurement of time that is available for use in other property definitions. Users may add to this units type. Declaration: Time_Units: **type units** (ps, ns => ps * 1000, us => ns * 1000, ms => us * 1000, sec => ms * 1000, min => sec * 60, hr => min * 60);

A.6 Runtime Services

Standard runtime services (as AADL subprograms) are defined in the AADL standard. The first set consists of service subprograms that can be called directly by an application source code. The second set consists of service subprograms that are callable by an AADL runtime executive and, generally, are not expected to be used directly by application component developers. An AADL runtime executive can be generated from an AADL model.

A.6.1 Application Runtime Services

The subprograms listed in Table A-12 may be called by application code.

Table A-12: *Application Runtime Services*

Property	Description

Send_Output	A non-blocking runtime service that allows a thread to explicitly cause events, event data, or data to be transmitted through outgoing ports to receiver ports. It takes a list of ports as an input parameter. This list specifies ports for which transmission is initiated. The send for all ports is considered to occur simultaneously. An exception is raised if the send fails with exception codes indicating the failing port and type of failure.
	
	 The *Send_Output* service replaces the *Raise_Event* service in the original AADL standard.
	
	 Declaration:
	   subprogram Send_Output
	   features
	    OutputPorts: **in parameter** <implementation-dependent port list>;
	    -- List of ports whose output is transferred
	    SendException: **out event data**; -- exception if send fails to complete
	    end Send_Output;

Table A-12: *Application Runtime Services (continued)*

Property	Description

Put_Value	A runtime service that allows the source text of a thread to supply a data value to a port variable. This data value is transmitted at the next *Send_Output* call in the source text or by the runtime system at completion time or deadline.
	
	 Declaration:
	   subprogram Put_Value
	   features
	     Portvariable: **requires data access**; -- reference to port variable
	     DataValue: **in parameter**; -- value to be stored
	     DataSize: **in parameter**; -- size in bytes (optional)
	   end Put_Value;

Receive_Input	A nonblocking runtime service that allows a thread to explicitly request input on its incoming ports to be frozen and made accessible through the port variables. The single input parameter specifies the list of ports for which the input is frozen. Any previous content of the port that has not been processed by *Next_Value* calls is overwritten. Newly arriving data may be queued. However, this does not affect the input to which the thread has access.
	
	 Declaration:
	   subprogram Receive_Input
	   features
	     InputPorts: **in parameter** <implementation-dependent port list>;
	     -- List of ports whose input is frozen
	   end Receive_Input;
	
	 For data ports the value is made available without requiring a *Next_Value* call. If a value has been updated, the service *Get_Count* returns the value 1 (i.e., it is fresh). If the data is not fresh, the value zero is returned.
	 For event data ports each data value is retrieved from the queue through the *Next_Value* call and made available as port variable value. Subsequent calls to *Get_Value* or direct access of the port variable will return this value until the next *Next_Value* call.
	 For event ports and event data ports the queue is available to the calling thread such that *Get_Count* returns the size of the queue. If the queue size is greater than one, the *Dequeued_Items* property and *Dequeue_Protocol* property may specify that more than one element is accessible to a thread.

continues

Table A-12: *Application Runtime Services (continued)*

Property	Description
Get_Value	A runtime service that allows a thread to access the current value of a port variable. The service call returns the data value. Repeated calls to *Get_Value* result in the same value returned to the caller, unless the current value is updated through a *Receive_Input* call or a *Next_Value* call. Declaration: **subprogram** Get_Value **features** Portvariable: **requires data access**; -- reference to port variable DataValue: **out parameter**; -- value being retrieved DataSize: **in parameter**; -- size in bytes (optional) **end** Get_Value;
Get_Count	A runtime service that allows a thread to determine whether a new data value is available on a port variable. For event port and event data port queues, the service provides the number of elements in the queue that are available to the thread. A count of zero indicates that no new data value is available. Declaration: **subprogram** Get_Count **features** Portvariable: **requires data access**; -- reference to port variable CountValue: **out parameter** BaseTypes::Integer; -- content count of port variable **end** Get_Count;
Next_Value	A runtime service that allows a thread to get access to the next queued element of a port variable as the current value. A *NoValue* exception is raised if no more values are available. Declaration: **subprogram** Next_Value **features** Portvariable: **requires data access**; -- reference to port variable DataValue: **out parameter**; -- value being retrieved DataSize: **in parameter**; -- size in bytes (optional) NoValue: **out event port**; -- exception if no value is available **end** Next_Value;

Table A-12: *Application Runtime Services (continued)*

Property	Description
Updated	A runtime service that allows a thread to determine whether input has been transmitted to a port since the last *Receive_Input* service call. The value true is returned if there is a new arrival. Declaration: **subprogram** Updated **features** Portvariable: **in parameter** <implementation-dependent port reference>; -- reference to port variable FreshFlag: **out parameter** BaseTypes::Boolean; -- true if new arrivals **end** Updated;

A.6.2 Runtime Executive Services

The service subprograms listed in Table A-13 may be called by application code or by an AADL runtime system that is implemented from an AADL model.

Table A-13: *Runtime Executive Services*

Property	Description
Get_Resource	The *Get_Resource* runtime service is provided to perform locking of resources using a specified concurrency control protocol. It may be used to lock multiple resources simultaneously. Declaration: **subprogram** Get_Resource **features** resource: **in parameter** <implementation-specific representation of one or more resources>; **end** Get_Resource;

continues

Table A-13: *Runtime Executive Services (continued)*

Property	Description
Release_Resource	The *Release_Resource* runtime service is provided to perform unlocking of resources using a specified concurrency control protocol. It may be used to lock multiple resources simultaneously. Declaration: **subprogram** Release_Resource **features** resource: **in parameter** <implementation-specific representation of one or more resources>; **end** Release_Resource;
Await_Dispatch	The *Await_Dispatch* runtime service is provided to support implementations of the runtime system that do not use a call-out mechanism to entrypoints (i.e., as a call-back mechanism in a cooperative scheduling implementation). A runtime service that suspends thread execution at completion of its dispatch execution. It is the point at which the next dispatch resumes. The service accepts a list of ports whose output is sent at completion or at deadline, a list of ports that potentially can trigger a dispatch, and a list of ports whose input is received at dispatch. The optional requires subprogram access *DispatchConditionFunction* provides a way of supplying an evaluation function to determine under what conditions a dispatch is triggered. The call involves several features. The in parameter *OutputPorts*, if present, is a list of ports whose sending is initiated at the completion of execution. This is equivalent to an implicit *Send_Output* service call. The in parameter *DispatchPorts* is a list of ports that potentially can trigger a dispatch. The out parameter *DispatchedPort* is a list of ports that did trigger a dispatch. The requires subprogram access feature *DispatchConditionFunction* is an optional function with a port list as a parameter that acts as a dispatch guard. This function identifies the ports, from those listed in *DispatchPorts*, that triggered a dispatch and the condition under which the dispatch triggered. If this function is not present, any of the ports in the *Dispatch-Ports* list can trigger the dispatch. The in parameter *InputPorts*, if present, is the list of ports whose content is received at dispatch. This is equivalent to an implicit *Receive_Input* service call.

Table A-13: *Runtime Executive Services (continued)*

Property	Description

Await_Dispatch *(continued)*	Declaration:
	  subprogram Await_Dispatch
	  features
	    -- list of ports whose output is sent at completion/deadline
	    OutputPorts: **in parameter** <implementation-defined port list>;
	    -- list of ports that can trigger a dispatch
	    DispatchPorts: **in parameter** <implementation-defined port list>;
	    -- list of ports that did trigger a dispatch
	    DispatchedPort: **out parameter** < implementation-defined port list>;
	    -- optional function as dispatch guard, takes port list as parameter
	    DispatchConditionFunction: **requires subprogram access**;
	    -- list of ports whose input is received at dispatch
	    InputPorts: **in parameter** <implementation-defined port list>;
	  end Await_Dispatch;

Raise_Error	A runtime service that allows a thread to explicitly raise a thread recoverable or thread unrecoverable error. An error type identifier is an input parameter.
	
	Declaration:
	  subprogram Raise_Error
	  features
	    errorID: **in parameter** <implementation-defined error type>;
	  end Raise_Error;

Get_Error_Code	A runtime service that allows a recover entrypoint to determine the type of error that caused the entrypoint to be invoked.
	
	Declaration:
	  subprogram Get_Error_Code
	  features
	    errorID: **out parameter** <implementation-defined error type>;
	  end Get_Error_Code;
	Since subprograms do not have an error port, if *Raise_Error* is called by a subprogram, the error is passed to the error port of the enclosing thread. If a *Raise_Error* is called by a remotely called subprogram, the error is passed to the error port of the thread executing the remotely called subprogram. *Raise_Error* has error identification as an output parameter value. This error identification can be passed through a thread's error port as a data value, since the error port is an event data port.

continues

Table A-13: *Runtime Executive Services (continued)*

Property	Description

Await_Result	A runtime service that allows an application to wait for the result of a subprogram call separate from a nonblocking call. This permits subprogram calls to execute concurrently.
	
	Declaration:
	  subprogram Await_Result
	  features
	    CallID: **in parameter** <implementation-defined call ID>;
	  end Await_Result;

Current_System_Mode	A runtime service that returns a value corresponding to the active system operation mode (SOM). This service assumes that system modes have identifying numbers.
	
	Declaration:
	  subprogram Current_System_Mode
	  features
	    ModeID: **out parameter** <implementor-specific>; -- ID of the mode
	  end Current_System_Mode;

Set_System_Mode	A runtime service that indicates the next system operation mode (SOM) to which the runtime system must switch. This service assumes that system modes have identifying numbers.
	
	Declaration:
	  subprogram Set_System_Mode
	  features
	    ModeID: **in parameter** <implementor-specific>; -- ID of the SOM
	  end Set_System_Mode;

Abort_Process	For a shutdown of a process due to an anomaly. All threads contained within the process are terminated without being given a chance to finalize. *Abort_Process* takes a process ID as a parameter.
Stop_Process	For a controlled shutdown of all threads contained within the process. All threads, whether executing, awaiting a dispatch, or not part of the current mode, are given a chance to execute their finalize entrypoint before the process is halted. This applies to all threads whether executing, awaiting a dispatch, or not part of the current mode. *Stop_Process* takes a process ID as parameter.

Table A-13: *Runtime Executive Services (continued)*

Property	Description
Abort_Virtual_ Processor	For a shutdown of a virtual processor due to an anomaly. All virtual processors and processes bound to the virtual processor are aborted. *Abort_Virtual_Processor* takes a virtual processr ID as parameter.
Stop_Virtual_ Processor	To initiate a transition to the virtual processor's stopping state at the next hyperperiod. This also initiates a *Stop_Virtual_Processor* for all virtual processors bound to the virtual processor, and a *Stop_Process* for all processes bound to the virtual processor. *Stop_Virtual_Processor* takes a virtual processor ID as parameter.
Abort_Processor	For a shutdown of a processor due to an anomaly. All virtual processors and processes bound to the processor are aborted. *Abort_Processor* takes a processor ID as parameter.
Stop_Processor	To initiate a transition to the processor's stopping state at the next hyperperiod. This also initiates a *Stop_Virtual_Processor* for all virtual processors bound to the processor and a *Stop_Process* for all processes bound to the processor. *Stop_Processor* takes a processor ID as parameter.
Abort_System	For a shutdown of the system due to an anomaly. All processors in the system are aborted.
Stop_System	To initiate a transition to the system stopping state. This will initiate a *Stop_Processor* for all processors in the system.

A.7 Powerboat Autopilot System

This is a summary of requirements for a powerboat autopilot (PBA). The PBA described here is illustrative and is provided for pedagogical purposes only. It is not intended for production nor does it, to the best knowledge of the authors, represent any product that is available or planned for manufacture or sale. Any resemblance to a commercial product is coincidental.

A.7.1 Description

The PBA system consists of the sensors, actuators, interface and display units, software, and computational hardware required to provide navigation and control of a powered water vessel. The PBA maintains a fixed speed and ensures that the vessel steers an appropriate course.

The PBA controller is a critical component in the PBA system. It uses current speed (from a speed sensor) and position information (from a GPS unit) to calculate appropriate commands to speed (e.g., throttle) and directional actuators (e.g., rudder or steering actuator for an outboard). In addition, the PBA controller receives input from a pilot interface unit and provides data to a pilot display unit.

The pilot's data inputs to the PBA control system are desired speed and standard operational parameters (e.g., future positions, desired speed). Some of the operational data are desired positions (waypoints). Up to five waypoints may be entered. One of these waypoints is designated as a terminal waypoint. When the terminal waypoint is reached, the PBA will gradually reduce the motor speed to idle and display the message 'arriving' to the user. During other times the display will reflect the status of the system including current speed, desired speed, current position, desired position (the next waypoint), and the system status (initialize, idle, control, off). There is a disengage signal that can be sent to the PBA control system by the pilot. This will terminate all control activity of the PBA.

There is a setup and maintenance (SAM) capability, where a user can enter and view configuration and control parameters. The pilot input and the pilot display unit are used as the interfaces for setup/maintenance. These may be integrated into a single unit (e.g., a handheld keypad with display that communicates with the PBA controller).

A.7.2 Enhanced Versions of the PBA System

The basic version of the PBA system has the capabilities outlined previously. Two other versions of the PBA product have enhanced capabilities and features and cost more than the basic. These are the deluxe and the prestige versions. The deluxe version has all of the capabilities of the basic plus a graceful degradation capability, an interface to depth and fish finding systems, extra operational features (e.g., multiple trips with more than five waypoints each), and high-resolution display capabilities. The prestige version, named the PBA super-system (PBASS), has all of the features of the deluxe version; however, the graceful degradation capability is replaced by continuous operation upon failure of the primary system. In addition, other sophisticated capabilities are included. (For example, the prestige version anticipates the vessel's arrival at the terminal waypoint and brings the boat to zero forward speed at the terminal waypoint.) Other capabilities of the prestige version include the ability to detect, produce a warning signal (audible and visual on the display), and avoid obstacles that are ahead in the water. Once a stationary object is detected, the autopilot will change the course of the powerboat, making a turn to starboard. Once past the object the autopilot will resume a direct course to the pending waypoint. If the detected object is moving, the autopilot will attempt to send out a message to the object, indicating the starboard turn. If confirmed, the boat will make the starboard turn. Otherwise, it will slow down and stop at a safe rate. This version also has capability to send a confirmation message if it receives a starboard turn message from another boat.

In the deluxe and prestige versions, there is an interface to a depth measuring and fish-locating device. This device is an intelligent device that provides the depth of the water below the vessel and location information for fish and related creatures in the water. The fish location data consists of GPS and depth coordinates. This data is sent to the autopilot with a period of 10hz for display on a console. In the deluxe and prestige versions setup can be done while the system is controlling the vessel. In the basic version, no automatic control is allowed during setup.

A.7.3 AADL Components of the PBA System

Figure A-1 presents a simplified AADL representation of the PBA system using AADL components and data and event port connections. We have made some architectural design decisions in creating this representation. First, since we do not intend to model the speed sensor, GPS unit, throttle, or directional actuator in detail, we represent them as devices. In contrast, we represent the pilot interface and display units as systems, allowing us the flexibility to develop detailed models of these components. We represent the PBA controller as an AADL process, reflecting the decision that the control functionality will be implemented in software.

We include the computational hardware resources (i.e., processor, memory, and bus) in the representation. Initially, we use a single processor 1GHz processor, a memory unit, and a bus. However, to achieve the capabilities of the other versions of the PBA, additional hardware resources (e.g., redundant processors) are required.

To simplify the presentation, we have included explicit data and event ports and connections, where the disengage signal is represented using event ports and connections. For all other transfers, we use data ports and sampling connections. We have not shown any physical access connections for any elements of the system.

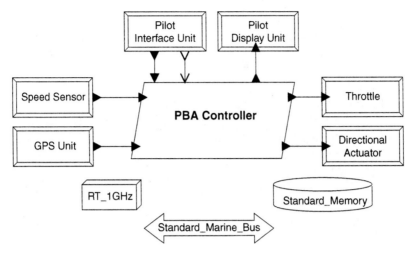

Figure A-1: *AADL Components for the PBA System*

A.7.4 An Alternative AADL Representation

In the initial representation, we used ports connections and explicit runtime categories in modeling the system. However, a generic representation that can serve as an architecture pattern (reference architecture) is shown in Figure A-2. This utilizes abstract components and feature groups. Through extensions of the elements of this reference architecture, a PBA specific model can be developed. For example, requisite ports and access features would be added to extensions of each of the feature groups.

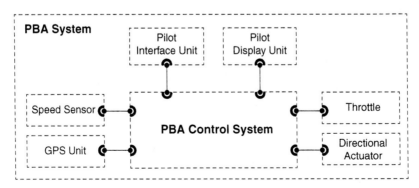

Figure A-2: *PBA System with Abstract Components*

Appendix B

Additional Resources

In this book, you have learned how to create architecture models of systems. Such models represent your software architectural design (i.e., the application source code organized into modules or packages as functional libraries or collections of classes and associated methods), your runtime architecture in terms of communicating tasks, your computer hardware architecture in terms of processors, memory, and buses (networks), and possibly a physical system with which your software system interfaces.

B.1 Modeling System Architectures

Here are some additional resources that can help you to become more proficient in modeling system architectures.

- Behjati, R., Yue, T., Nejati, S., Briand, L., and Selic, B. An AADL-Based SysML Profile for Architecture Level Systems Engineering: Approach, Metamodels, and Experiments, Simula Research Laboratory, ModelME Technical Report 2011-03, Feb 2011.

  This report discusses a UML profile that extends SysML with AADL concepts, which was implemented in Rhapsody.

- Feiler, P. H. *Modeling of System Families* (CMU/SEI-2007-TN-047). Pittsburgh, PA: Software Engineering Institute, Carnegie Mellon University, 2007.

This report discusses how you can go about representing a family of systems in AADL. It discusses multiple dimensions of variability in such a system family and how each is best expressed in AADL.

- Feiler, P. H. and Hansson, J. *Flow Latency Analysis with the Architecture Analysis and Design Language (AADL)* (CMU/SEI-2007-TN-010). Pittsburgh, PA: Software Engineering Institute, Carnegie Mellon University, 2007.

  This report goes into depth in modeling end-to-end flows and performing end-to-end latency analysis. This analysis can provide a lower bound for worst-case latency at different levels of fidelity. It takes into account a range of runtime system contributors to latency and latency jitter.

- Feiler, P. H. and Hansson, J. *Impact of Runtime Architectures on Control System Stability*, 4th International Embedded Real-Time Systems Conference (ERTS), Toulouse, France, Jan 2008.

  This article discusses the criticality of managing end-to-end latency jitter by minimizing non-deterministic sampling of the data stream, as such jitter potentially leads to controller instability.

- Feiler, P. H. and Rugina, A. *Dependability Modeling with the Architecture Analysis and Design Language (AADL)* (CMU/SEI-2007-TN-043). Pittsburgh, PA: Software Engineering Institute, Carnegie Mellon University, 2007.

  This report introduces you to the Error Model Annex standard for AADL and its use to capture intrinsic fault behavior as well as fault propagation behavior of your system, such that stochastic analytical models can be generated for reliability and availability analysis as well as fault tree analysis.

- Hudak, J. and Feiler, P.H. *Developing AADL Models for Control Systems: A Practitioner's Guide* (CMU/SEI-2007-TR-014). Pittsburgh, PA: Software Engineering Institute, Carnegie Mellon University, 2007.

  This report introduces you to creating a control system model.

B.2 Cases Studies

The authors and other members of the SEI AADL team have performed a number of case studies in modeling application systems.

- Barott, J. B., Gluch, D. P., and Kirby, S. L. "Predictive engineering of an unmanned aerial system (UAS) using the Architecture Analysis and Design Language (AADL)," *Systems Conference (SysCon), 2011 IEEE International*, pp. 569–573, 4–7 April 2011.

  This paper presents the results of applying predictive software-dependent system engineering practices using the SAE International Architectural Analysis and Design Language (AADL) in the modeling and analysis of an unmanned aerial system (UAS), part of a search and rescue (SAR) system. The SAR system embodies many of the challenges associated with engineering complex software-dependent systems, such as achieving stringent performance requirements and ensuring effective resource utilization. The results of this work demonstrated that model-based software system engineering practices employing the AADL can be used to analyze important system aspects early in architectural development, and can be an integral element in making informed decisions throughout an engineering effort

- Feiler, P. H., Lewis, B. (US Army AMCOM), and Vestal, S. (Honeywell Technology Center). *Improving Predictability in Embedded Real-Time Systems*, Special Report CMU/SEI-2000-SR-011 December 2000.

  This paper discusses an early case study in using an architecture modeling language for embedded systems to improve predictability of performance in embedded real-time systems. This approach uses MetaH, the language AADL is based on. It utilizes automated analysis of task and communication architectures to provide insight into schedulability and reliability during design and automatic code generation to produce a complete system. The approach has been applied to a missile guidance system. A port from a single processor system to a dual processor system was accomplished in three weeks and the missile flew correctly in a simulation test.

- Feiler, P. H., Gluch, D., Hudak, J., and Lewis, B. *Embedded System Architecture Analysis using SAE AADL*, Technical Note CMU/SEI-2004-TN-005, 2004.

This report discusses a pattern-based approach we used to analyze an avionics architecture that is migrating from a federated approach to an integrated modular avionics (IMA) approach.

- Feiler, P. H., Hansson, J., de Niz, D., and Wrage, L. *System Architecture Virtual Integration: An Industrial Case Study*, Technical Report CMU/SEI-2009-TR-017, November 2009.

This report discusses the proof of concept phase of an aerospace industry initiative to improve its practice to become more technology intensive and architecture centric. Under the umbrella of the Aerospace Vehicle Systems Institute (AVSI) the System Architecture Virtual Integration project in this first phase has defined the to-be process, a return on investment (ROI) model, and a proof of concept demonstration of architecture modeling and analysis of multiple quality attributes at different levels of fidelity early and throughout the development process. The demonstration scenario included the use of a model repository and model bus starting with an aircraft model to illustrate dealing with system engineering and embedded software system engineering as well as supporting model-based interaction between system integrators and suppliers.

- Feiler, P. H., Gluch, D. P., and Woodham, K. *Case Study: Model-based Analysis of the Mission Data System Reference Architecture*, Technical Report CMU/SEI-2010-TR-003, May 2010.

This report presents the results of a case study applying the Architecture Analysis & Design Language (AADL) to the Mission Data System (MDS) architecture. This work is part of the NASA Software Assurance Research Program (SARP) research project: "Model-Based Software Assurance with the SAE Architecture Analysis & Design Language (AADL)." In this report, we discuss modeling and analyzing the MDS reference architecture. In particular, we focus on modeling aspects of state-based system behavior in MDS for quantitative analysis. Three different types of state-based system model are being considered: closed loop control, goal-oriented plan execution, and fault tolerance through replanning.

- Hansson, J., Lewis, B., Hugues, J., Wrage, L., Feiler, P., and Morley, J. *Model-Based Verification of Security and Non-Functional Behavior using AADL*, IEEE Security & Privacy, Page: 1-1, Jan 2010.

Modeling of system quality attributes, including security, is often done with low fidelity software models and disjointed architectural specifications by various engineers using their own specialized notations. These models are typically not maintained or documented throughout the life cycle and make it difficult to obtain a system view. However, a single-source architecture model of the system that is annotated with analysis-specific information allows changes to the architecture to be reflected in the various analysis models with little effort. We describe how model-based development using the Architecture Analysis and Design Language (AADL) and compatible analysis tools provides the platform for multidimensional, multifidelity analysis, and verification. A special emphasis is given to analysis approaches using Bell-LaPadula, Biba, and MILS approaches to security and that enable a system designer to exercise various architectural design options for confidentiality and data integrity prior to system realization.

For additional papers on the use of AADL by the community, please see the AADL Related Publications section of the public AADL Wiki [AADL Web].

Appendix C

References

All URLs are valid as of the publication date of this document.

AADL Web

The Architecture Analysis & Design Language (AADL) Information Site (http://www.aadl.info) and public Wiki (https://wiki.sei.cmu.edu/aadl).

SEI Technology Highlight: AADL and Model-based Engineering (http://www.sei.cmu.edu/library/assets/ResearchandTechnology_AADLandMBE.pdf).

AADLSimulink

Gopal Raghav, Swaminathan Gopalswamy, Karthikeyan Radhakrishnan, Jérôme Hugues, Julien Delange, Model Based Code Generation for Distributed Embedded Systems Embedded Real-time Software and Systems Conference (ERTS2010), May 2010.

Ariane 5

"Ariane 5 Flight 501." en.wikipedia.org/wiki/Ariane_5_Flight_501.

AS5506A

"Architecture Analysis & Design Language (AADL)," SAE International Standards document AS5506A, Nov 2004, Revised Jan 2009. http://www.sae.org/technical/standards/AS5506A.

AS5506/1

"Architecture Analysis & Design Language (AADL) Annex Volume 1," SAE International Standards: AS5506/1, 2006. http://www.sae.org/technical/standards/AS5506/1.

AS5506/2

"Architecture Analysis & Design Language (AADL) Annex Volume 2," SAE International Standards: AS5506/2, 2011. http://www.sae.org/technical/standards/AS5506/2.

ASIIST

Application Specific I/O Integration Support Tool. University of Illinois. https://agora.cs.illinois.edu/display/realTimeSystems/ASIIST.

BAnnex

Architecture Analysis & Design Language (AADL) Annex Volume 2: Annex D: Behavior Annex, SAE International Standards: AS5506/2, 2011. http://www.sae.org/technical/standards/AS5506/2.

BLESS

Behavioral Language for Embedded Systems with Software. Draft Annex standard. Brain Larson, brl@multitude.net.

BOE

Boeing 777 Family of Aircraft. http://www.boeing.com/commercial/777family/background.html.

CAT

Consumption Analysis Toolbox. https://sourceforge.net/projects/lab-sticc/files/CAT/.

Casteres 09

J. Casteres and T. Ramaherirariny, "Aircraft integration real-time simulator Modeling with AADL for architecture tradeoffs," Proceedings of 9th Design, Automation & Test in Europe Conference, April 2009.

Cheddar

Real-time Scheduling Analyzer. http://beru.univ-brest.fr/~singhoff/cheddar/.

Clements 10

P. Clements, F. Bachmann, L. Bass, D. Garlan, J. Ivers, R. Little, R. Nord, and J. Stafford. *Documenting Software Architectures: Views and Beyond.* Second Edition. Boston, MA: Addison-Wesley, 2010.

COMPASS

Correctness, Modeling and Performance of Aerospace Systems (COMPASS) project and tool chain. http://compass.informatik.rwth-aachen.de/.

DAnnex

Architecture Analysis & Design Language (AADL) Annex Volume 2: Annex B: Data Modeling Annex, SAE International Standards: AS5506/2, 2011. http://www.sae.org/technical/standards/AS5506/2.

DDS

OMG Data Distribution Service. http://www.omgwiki.org/dds.

DeNiz 06

D. de Niz and R. Rajkumar, "Partitioning bin-packing algorithms for distributed real-time systems," *International Journal of Embedded Systems,* vol. 2, no. 3-4, pp. 196–208, 2006.

DeNiz 08

Dio Deniz and Peter Feiler, "On Resource Allocation in Architectural Models," Proceedings of the 11th IEEE International Symposium on Object Oriented Real-time Distributed Computing (ISORC), May 2008.

EAnnex

Architecture Analysis & Design Language (AADL) Annex Volume 1: Annex E: Error Model Annex, SAE International Standards: AS5506/1, 2006. http://www.sae.org/technical/standards/AS5506/1.

Eclipse

www.eclipse.org.

Feiler 07

P.H. Feiler and A. Rugina, "Dependability Modeling with the Architecture Analysis & Design Language (AADL)" (CMU/SEI-2007-TN-043). Pittsburgh, PA: Software Engineering Institute, Carnegie Mellon University, 2007. http://www.sei.cmu.edu/library/abstracts/reports/07tn043.cfm.

Feiler 07A

P.H. Feiler, "Modeling of System Families," (CMU/SEI-2007-TN-047). Pittsburgh, PA: Software Engineering Institute, Carnegie Mellon University, 2007. http://www.sei.cmu.edu/library/abstracts/reports/07tn047.cfm.

FIACRE

B. Berthomieu, et. al., "FIACRE: an Intermediate Language for Model Verification in the TOPCASED Environment," Proceedings of 4th International Congress on Embedded Real-Time Systems, Jan 2008. AADL to FIACRE Translator, http://gforge.enseeiht.fr/projects/aadl2fiacre.

Hansson 08

J. Hansson and P.H. Feiler, "Enforcement of Quality Attributes for Net-centric Systems through Modeling and Validation with Architecture

Description Languages," Proceedings of 4th International Congress on Embedded Real-Time Systems, Jan 2008.

Hofmeister 00

C. Hofmeister, R. Nord, and D. Soni, *Applied Software Architecture*, Object Technology Series. Boston, MA: Addison-Wesley, 2000.

IEEE 42010

Institute of Electrical and Electronics Engineers. Systems and software engineering—Architecture descriptions (IEEE Std 42010-2011). New York, NY: Institute of Electrical and Electronics Engineers, 2011.

MARTE

Modeling and Analysis of Real-time and Embedded systems. Object Management Group (OMG). http://www.omgmarte.org/.

NASA

P. Feiler, D. Gluch, K. Weiss, and K. Woodham, "Model-Based Software Quality Assurance with the Architecture Analysis and Design Language," Proceedings of AIAA Infotech @Aerospace 2009 (April 2009).

Ocarina

Ocarina: Code Generator for AADL. http://ocarina.enst.fr/.

MDA

Model Driven Architecture. Object Management Group (OMG). http://www.omg.org/mda/.

OSATE

The Open Source AADL Tool Environment (OSATE) for the SAE Architecture Analysis & Design Language (AADL). http://www.aadl.info.

RC META

The Rockwell Collins META toolset. https://wiki.sei.cmu.edu/aadl/index.php/RC_META.

SAVI

David Redman, Donald Ward, John Chilenski, and Greg Pollari, "Virtual Integration for Improved System Design," Proceedings of The First Analytic Virtual Integration of Cyber-Physical Systems Workshop in conjunction with RTSS 2010.

See also, P. Feiler, L. Wrage, and J. Hansson, "System Architecture Virtual Integration: A Case Study," Embedded Real-time Software and Systems Conference (ERTS2010), May 2010.

STOOD

A tool environment for HOOD-RT and AADL used in European Aerospace industry. http://www.ellidiss.com.

SysML

Systems Modeling Language (SysML). Object Management Group (OMG). http://www.omgsysml.org/.

TASTE

TASTE (The ASSERT Set of Tools for Engineering) project and tool chain led by the European Space Agency. http://assert-project.net/taste.

TOPCASED

TOPCASED: The Open Source Tool for Critical Systems. http://www.topcased.org/.

UML

Unified Modeling Language (UML). Object Management Group (OMG). http://www.omg.org/technology/documents/modeling_spec_catalog.htm#UML.

W3C 04

World Wide Web Consortium (W3C). *Extensible Markup Language (XML) 1.0 (Third Edition).* http://www.w3.org/TR/2004/REC-xml-20040204/ (2004).

Index

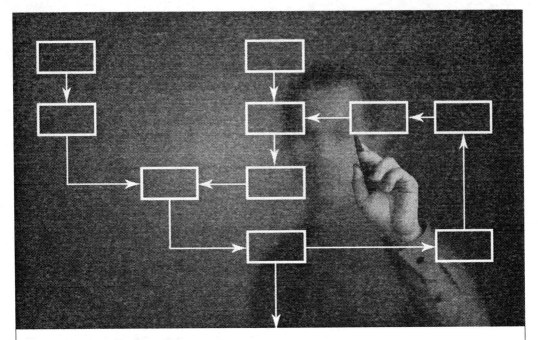

Learn to Apply Model-Based Engineering Concepts with Training from the SEI

Do you need to resolve application scheduling and resource conflicts? Are you confident that your control-system applications will perform after a move to a new platform? Is your development practice robbing you of productivity?

In the 4.5-day SEI Modeling System Architectures Using the Architecture Analysis and Design Language (AADL) course, you will receive a solid overview of system and software modeling and evaluate your options for engineering embedded real-time systems. Course exercises will provide you with a concrete context for the issues.

You will learn how to use tools for model-based engineering that can
- reduce risk through early analysis of system architecture
- reduce cost through fewer system-integration problems
- assess system-wide impacts of architectural choices
- validate modeling assumptions in the operational system

For more information about the SEI Modeling System Architectures Using the AADL course, visit www.sei.cmu.edu/goto/aadl-course.

For information about SEI software architecture credentials, visit www.sei.cmu.edu/go/architecture-credentials.